SCIENCE, MATH AND NUTRITION
for Toddlers

SCIENCE, MATH AND NUTRITION
for Toddlers

Setting the Stage for Serendipity

— by —

Rita Schrank

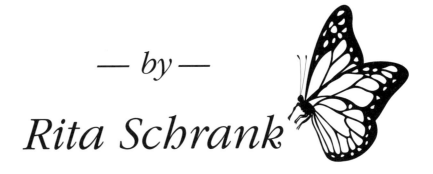

Includes:

Reproducible

PARENT PAGES

ANNOTATED BIBLIOGRAPHIES:

Adult Resource Books

Early Childhood Books

Cassettes, CD's, Videos & Toys

HOMESCHOOL INFORMATION

— Humanics —

SCIENCE, MATH AND NUTRITION FOR TODDLERS:
SETTING THE STAGE FOR SERENDIPITY
A Humanics Publication

Published in the United States

Humanics Learning is a division of Humanics Limited.
Humanics Limited, P.O. Box 7400, Atlanta, GA 30309

PRINTED IN THE UNITED STATES OF AMERICA

ISBN 0-89334-280-7

Library of Congress Cataloging-in-Publication Data

Schrank, Rita.
        Science, math and nutrition for toddlers: setting the stage for serendipity/ by Rita
Schrank.
            p.   cm.
        Includes bibliographical references and index.
        ISBN 0-89334-283-1. -- ISBN 0-89334-280-7 (pbk.)
        1. Science--Study and teaching (Preschool)--Handbooks, manuals, etc. 2. Mathematics--Study and teaching (Preschool)--Handbooks, manuals, etc. 3. Nutrition--Study and teaching (Preschool)--Handbooks, manuals, etc. 4. Toddlers--Education (Preschool)--Handbooks, manuals, etc.
I. Title.
LB1140.5.S35S37 1998
372.21--dc21   98-23246
                                                                                        CIP

To my husband Glen
who introduced us all to the joys of learning
and the beauties of nature.

Contents

MATH ACTIVITIES

"The fairest thing we can experience is the mysterious.
It is the fundamental emotion which stands at the cradle of true art
and true science. He who knows it not and can no longer wonder,
no longer feel amazement, is as good as dead."

Albert Einstein
The World As I See It

"The principle art of the teacher is to awaken the joy in creation & knowledge."

Einstein, The World As I See It

Introduction: The Adult Role

Richard Feynman, the famous physicist, tells of his father's role in his becoming a scientist. His father willingly answered the "whys," not necessarily with answers, but with questions...intriguing questions. The elder Feynman shared his love of "finding out things" with his son. As a consequence, Dr. Feynman went through his life loving to solve the mysteries of "what if..." and "How do you suppose...."

Including science, math and nutrition in a toddler's learning can start now, and doing so will be fun, but be forewarned that the best fun will come between the child's ages of 4 and 13. As hard as it will be, as the adult you will need to SAVE many of those tantalizing science projects until the child is older. You can do a great deal of foundation building right now however. The challenge is to keep things simple. Just know that you are witnessing some of the most dramatic influential changes in all of human growth and that you have an opportunity to guide, challenge and excite a toddler's curiosity through this special time. Approach with joy and caution.

During my research I found many wonderful ideas and books that were a little old for toddlers. These are listed in the appendix and can be used as resources and/or be "talked" with toddlers until they grow into the books.

From their earliest years children encounter an infinite number of math, science and nutrition experiences in daily life. They react to ice melting, seasonal changes, sun, moon, stars, clouds, bugs, plants, friction and inertia. So it should not be difficult to start recognizing and savoring the most simple experiences with toddlers. You will soon note that most experiences are not divided into neat subjects (physics, math) the way schools divide them.

The adult's role in a toddler's learning is that of "research assistant." During the earlier, non-verbal years of toddlerhood the adult will need to be a partner-in-learning. Then as the toddler grows, the adult's role becomes less and less, until all that will be needed from the adult is the role of facilitator or an audience.

Curiosity, questioning, seeking, and open-mindedness are some basic characteristics of scientific inquiry. Toddlers fit this bill very well indeed. The adult can provide a rich environment (set the stage), observe toddlers, encourage their natural curiosity, ask thought-provoking questions, answer so that further questions are born, encourage an attitude of "let's try," and be willing to change a plan or an opinion, based on the new evidence. It is important that the toddler be free to experiment and explore. All of this, including real

life problem solving, is a part of the development of scientific thought. Above all, adults must try reach back into their childhoods, and share that sense of wonder and amazement as they and toddlers enjoy the world around them. Joseph Cornell, who wrote *Sharing Nature with Children,* gives a few suggestions for good teaching: Look and experience first, talk later. Be receptive, listen more. Share more and teach less. Enthusiasm is contagious—so delight in it with toddlers! Let toddlers lead you and enjoy this special time.

What is a Toddler?

A toddler is not an infant, yet he is too young to be a preschooler. Just when you think that you are getting good at managing an infant, a dramatic change takes place. Before you know it, the child is no longer an a tiny baby. All of a sudden your baby gets up on his two feet and begins to have a mind of his own. You guessed it—the infant is now a toddler.

Before you begin to think you've done something wrong—worrying that maybe you are caring for a potential delinquent—let me introduce you to toddlers.

They are not yet verbal preschoolers and "reasoning" with them will not yet work. They are still learning by instinct and context, and have a way to go before they can understand sharing and "being nice" or learn to read.

Toddlers are special. Toddlerhood—a stage, not an age, is a time when children's' rudimentary curiosity is high and is spurred on by their rapidly developing coordination. They plunge headlong into life, uninhibited, ready to taste and test every inch of this new world. The window of opportunity for learning the foundations of science, math and nutrition is wide open.

Scientist

No one fits the description of scientist better than a toddler, who is non-judgementally testing and retesting the world around with no preconceived notions, ready to come to conclusions about the world's nature. After all, she is seeing it for the very first time.

A toddler just loves to pick up something, feel it, taste it and drop it, over and over and over again, just to be sure that it really always does what it is doing. "Is that the way it really does work?"

How Toddlers Learn

Toddlers learn by watching and helping and by doing things themselves. They need to explore. Toddlers need to talk about everything. Although they need novelty and simplicity, they love repetition.

Helping

There are many things that can be learned by what we call "partnership learning." Most of the activities in the following pages are planned as partnership exercises. Just as a baby learns to take off a sock by having it pulled part way, so a toddler learns by being a part of the team. Young toddlers watch a lot and do those parts of a project that they can do. As competency grows, toddlers do more and more, and the adults do less and less.

Doing Things Themselves

Try to let your toddler do as much by himself as possible. Remember, some struggle is a part of learning.

Exploring

Here is where a toddler shines. A child does not need teaching or "help" to attain the pre-programmed stages of sensory motor and gross motor development. These are best learned while the toddler is freely and independently exploring and manipulating in a safe, planned environment You set the stage and the process itself provides the pleasure. Then it is up to you to watch for those teach-able moments" in which to extend and enrich the learning.

Talk About It

Learning language is the most important job at this age. Give a toddler the gift of words—not lectures. Help label things, ideas, and feelings at every opportunity, extending and enriching toddler"s language, and con-necting the words to concrete objects when-ever possible. If a toddler says, "dog" you can say, "Yes, a big, black dog."

Novelty and Simplicity

Provide variety, but not confusion. The toddler may come upon an elaborate collage art project in a nursery school and will delightedly pour the glue, crumple or eat the pretty materials and clear everything off the table to the floor. The child is not being "naughty" but is doing what toddlers like to do: pour, dump, and watch gravity at work. It is possible to do "projects" with a very young toddler. "One step at a time" should be a motto for working with toddler. Keep all the materials out of reach until it is time to introduce them.

Incidentally, this method of introducing one simple step at a time helps toddlers lengthen their attention span.

Thoughts About Magic

Some fascinating science projects are done to amaze older children. Toddlers, however, cannot be amazed. The toddler's whole world is magic and whatever happens is "how things are." Therefore doing magical things now is wasted for the moment. Wait until the child understands some of the rules of the world, like gravity, sun rises, moon rises, etc...your four-year-old is ready to be amazed.

Repetition

Toddlers love to have adults read something to them over and over again—it can drive a person crazy—but that's the way toddlers are. As we have said, they love repetition—prac-tice, practice, practice! Some play and learn-ing materials provide opportunities to learn "science" or "math" each time the child plays with them. Water, sand, playdough, blocks and cooking seldom grow dull to toddlers. These things will still be favorites of kinder-gartners and first graders.

Laying Foundations

The toddler's world is fresh, new and beautiful. It is up to us to rediscover this world for ourselves, share it with them and help them establish the groundwork for future learning.

Prologue: Learning and Intelligence

Each of us arrives on this earth with a different set of potentials—both strengths and weaknesses. We are obliged to play the cards that nature has dealt us. The various standardized intelligence and IQ tests and other social and developmental assessments that are in use are designed to suggest what hand the child has been dealt. Then, knowing the child's strengths and weaknesses, we can design our approach to helping the child develop to his maximum potential.

As we work with toddlers, many early childhood people and I appreciate Professor Howard Gardner's theory of seven intelligences:

1. Musical ("music smart")
2. Linguistic ("word smart")
3. Logical/mathematical ("logic smart")
4. Bodily/kinesthetic ("body smart")
5. Visual/spatial ("picture/image smart")
6. Knowledge of self—intrapersonal ("self smart")
7. Knowledge of others-interpersonal ("people smart").

I like his theory's simplicity and the fact that Gardner says that it is "just a theory to try for a while" *and* Gardner values intelligences other than the kinesthetic and logical "smarts." The world would be bereft indeed if we were all Michael Jordans or Albert Einsteins. This gives us seven areas in which to build a foundation for future learning by toddlers.

Because we are focusing on science and math we will of course be emphasizing the mathematical/logical, linguistic and visual/spatial activities more than interpersonal and intrapersonal activities. However, let's remember that toddlers learn with their whole selves, spiced and energized by their emotions.

MUSIC & THE BRAIN

Music, one of the first modes of learning, should be your special tool as you help toddler learn about the world around them. As a toddler is exposed to music's patterns, rhythms, symbols and language, the activity helps fine tune the brain's circuitry. Music helps integrate both sides of the brain, stimulating them to work together. It is known that learning is retained more efficiently when both sides of the brain are active. The left side of the brain is logical and math organized; this part of the brain helps us read and write music, but cannot recognize a melody. The right side uses intuition and recognizes patterns in art, music and in life. From infancy each stimulating experience aids the miraculous process of wiring young brains for a lifetime of learning, and music can provide many of those stimulating experiences.

Only a few months ago our toddlers graduated from merely looking about to being able to reach and grasp. Now on their feet as young toddlers they begin to experiment with simple cause and effect experiments, explore places and examine things and by age two they will move into a new stage of development when—no longer limited to sensory/motor learning—they begin to develop concepts. They still rely heavily upon sensor/motor experiences, but can now sock away ideas into their thought processes and begin to express these ideas to others—an exciting two-year adventure, indeed. Then success, satisfied curiosity, and lovely sensations like having mud squeeze through their toes, contribute to the 730 days of astounding growth called toddlerhood, which promises to produce an amazing 3-year-old. At age one a toddler speaks one or two words, and at three her vocabulary includes up to 1000 words.

In the Appendix you will find a modest description of ages and stages to help you have more realistic expectations of your tod-dler. Then you can plan "developmentally appropriate" activities as you identify windows of opportunity for learning. We hope this outline will help you avoid intimidating a toddler or avoid massive adult intrusion when toddler should be left alone. Try to practice academic freedom. The only way a child can develop the ability to concentrate is to practice concentrating—so try not to interrupt an involved toddler.

Toddlers learn best when they are allowed to enjoy the exquisite pleasure of freely exploring and manipulating in a safe setting. Sometimes, in the interest of the child's development, we will want to carefully design the environment and set the stage. But, because toddlers are in what famed Swiss psychologist Jean Piaget calls the sensorimotor stage, they "know" the world exclusively through their own perceptions and actions upon it. So most of the time, concrete and relevant "real life" experiences will provide better opportunities for toddler learning. Either way, as we facilitate a toddler's learning, we can keep Gardner's seven basic "smarts" in mind, challenge both sides of the brain to work together, and "feed all the child's senses" as we build that dynamic foundation for future learning.

Parent Pages

"If facts are the seeds that later produce knowledge & wisdom,
then the emotions & the impressions of the senses are the fertile soil
in which the seeds must grow."

Rachel Carson, Sense of Wonder

Reader's Notes

6/4 - ¹Let Paul wash out ice tray after each use. ²Let him watch filling the little tub w/ soapy water. ³Let him dip plastic & silverware in tub and rinse / dry these items.

4. Let him count tsp of "dip" you place in tray and describe the texture: yogurt - smooth, thick, etc.

5. Assess the toys in playroom, Paul's & Phoebe's im - are these still age appropriate?

6 Recycling - Let Paul wash out yogurt cups and place in recycling bag.

"Parents possess a lot more knowledge than they think they do. Help them share that knowledge with their children."

Introduction to the Parent Pages

You hear that the home-school connection needs to be strengthened—and as early in the child's life as possible. We hope the Parent Pages will help you do this.

We all know that a parent is the child's first teacher. However, what is being learned at home is not always respected or labeled as learning, let alone labeled as "science" or "math" or "nutrition." So herewith are the Parent Pages—links with home that we hope will help narrow the parent information gap about children's learning, bolster parents' respect of their own knowledge, and help them enjoy sharing that knowledge with their children. Parents know a lot about the world that toddlers need to know.

The reproducible Parent Pages are intended to promote parent-child interactions, enhance school-based learning (without becoming a burden on parents), and initiate real-world connections of science, math and nutrition concepts.

Help parents focus on and respect an ordi-nary event as a learning opportunity. Encourage them to enjoy their time with their children. Suggest that they go out with their kids, explore their environment and learn together. Encourage them to observe, label, sort, count and communicate, along with their enjoyment. Then it will be your job to make an opportunity for toddlers to share what they can about those family excursions in order to reinforce that learning.

Parents At School

Helping out at school gives parents a feel-ing that they are important in their child's education. T. Berry Brazelton said, "Rather than simply set up educational opportunities for parenting or advice-giving groups, we need to give young parents the feeling that they are important, that they have choices and can make them appropriately." Give par-ent well-defined tasks and let them know they are appreciated.

Science

Setting the Stage for Learning

"The whole of science is nothing more than a refinement of everyday thinking." —*Einstein*

Dear Parents of Toddlers:

You don't need to be an Einstein to help your child learn about science. Toddlers cannot be *taught* science. You can, however, take advantage of daily experiences to prepare toddlers to understand scientific concepts and to use the scientific method for learning. Toddlers are already better scientists than we are. They enthusiastically taste and test their world without any preconceived notions. You can help toddlers keep that glorious sense of wonder and help them learn basic science at the same time. You know plenty of science: you know that ice melts, that clouds bring rain, and that if we don't water plants they die. If you think about it, you know that cooking is chemistry and physics, as foods change form, and also math, as you follow recipes. Our lives are increasing driven by science and technology and you are a part of that world. Share it with your toddler. Learn together and enjoy.

OUTSIDE

Go for a Walk: Next time you go for a walk with your kids, think about what you and your child are learning. Sharpen your skills of observation and talk about what you see and hear and smell. If you find a bug or a item that you cannot identify, say, "Lets find out."

INSIDE

In the Kitchen: Cook together—your toddler will have to watch a lot and "help" a little. Talk about how foods change—they freeze, melt, boil, thicken.

In the Bathroom: The water is cold or hot or just right. Water pours and splashes. The toilet flushes. Some things float, some things do not.

In the Living room: The TV is loud, not loud enough. Describe what is happening on TV (i.e., the football player is running fast). Find a children's show that you can watch together.

In the Washroom: The soap dissolves in the water, the clothes get wet, then hot, then they get dry. (Have you ever dried with a towel still warm from the dryer?)

Whatever you do with your toddler can be an exciting learning experience for him. After all he is new to this world. Your job is to keep your eye open for opportunities to help toddler learn something new—a new word, a new idea, a label. A non-verbal toddler will especially appreciate new words.

We don't want to teach science to toddler, but to expose her to a way of thinking about the world—one where she sees the world in a special way, asks questions, and finds out.

Joseph Cornell, who wrote *Sharing Nature with Children,* gives a few suggestions for good teaching: **Look and experience first**—talk later. Be receptive—listen more. Teach less and share more. Enthusiasm is contagious—so delight in it with toddlers! Let toddlers lead you and enjoy.

NOTE: As you share everyday experiences with children, you will be building lifetime memories. Find and enjoy Tom Hunter's cassette tape *Memories,* made with Bev Bos and Michael Leeman.

Enjoy Chubby English/Spanish Books, *Let's Take a Walk; Vamos a Caminar.* NY Little Simon Merch. For Parents of older children: *Helping Your Child Learn Science,* Wash. D.C. GPO 1994.

From: *Science, Math & Nutrition for Toddlers: Setting the Stage for Serendipity* by Rita Schrank, Humanics, Atlanta, GA

Learning Math Concepts

Dear Parents of Toddlers:

No one can *teach* toddlers math. However, even if you are not "good" at math, you can take advantage of daily experiences to get toddlers ready to understand math concepts. Toddlers are enthusiastically tasting and testing their new world. You can help them keep that glorious sense of wonder and help them enjoy learning some math basics.

DOING REAL THINGS TOGETHER

You and your toddler can take advantage of real life activities to count (for example, how many sandwiches you need), sort (socks or clean clothes), and categorize (canned goods or toys) together.

SEQUENCES AND PATTERNS—both math concepts—are all around us:

Daily Rituals: "First we use the toilet, then we wash our hands and then we brush our teeth. After that we will read two books, then it will be time to sleep."

Nursery rhymes, clapping games, songs or just having fun with numbers and words are all good ways to expose a toddler to math concepts:

"One two; buckle my shoe
Three four; shut the door,
Five six; pick up sticks…"

"See you later, alligator—
After while, crocodile."

"One Potato, two potato, three potato four. Five potato, six potato, seven potato more."

Also Raffi's *Time to Sing, Rise and Shine* and *Shake My Sillieses Out* are great beginning math concept songs.

> "Waiting is the hardest thing that a toddler needs to do."

WAITING

Waiting is the hardest thing that a toddler needs to do. Children are not very aware of time, so let's take advantage of waiting to expose toddler to the ideas of sequencing: first we do this, then we do that. If you have a verbal toddler, *PREPARE* her ahead of an anticipated wait. Say, "When we go to the doctor's we will say hello to the nurse, and then we will wait. While we wait you can read a magazine if you like. The nurse will call us when it is our turn. Then we will talk to the doctor." A non-verbal toddler will understand, at least, the context of what you say especially if you "talk her through" the waiting steps. Of course, some very young toddlers' bodies will need to move even while waiting. A short walk until it is their turn may be best.

NOTE: I spy with my little eye is a good waiting game for older children.

From: *Science, Math & Nutrition for Toddlers: Setting the Stage for Serendipity* by Rita Schrank, Humanics, Atlanta, GA

Learning About Healthy Foods

Science, Math and Nutrition

Dear Parents of Toddlers:

Cooking and eating are great ways to learn about science, math and nutrition. Toddlers will enjoy watching you cook and can even help in small ways—they love to wash anything and especially like to stir. As the toddler grows he can help more and more.

Bit by bit you and your toddler will learn to enjoy socializing with family and friends. Shared meals and snacks offer people a way to express affection and nourish the mind as well as the body. Make shared eating times with your toddler short at first.

> *Healthy food habits are "caught" more than taught. Toddlers need models, not critics.*

Science: The kitchen is a laboratory where chemistry, physics and math surround us. Heat changes the food. Ice melts. Eggs change in a number of different ways. Foods change their color and their taste as they cook. Dry ingredients mixed with wet ones turn into magical things like cakes. Look at your cooking in a scientific way—talk about it with your toddler and enjoy it.

Math: Measuring ingredients and planning numbers of servings or how long to cook and setting the table offer opportunities to introduce real-life math ideas to the toddler.

Nutrition: Knowledge about nutrition is "caught," not necessarily taught. We admit that what we and our children *will* eat and what we *should* eat are not necessarily the same. The best we can do is follow these suggestions:

1. Good food is good whenever it is eaten. Toddler's stomachs are small and they probably need at least five small meals each day—peanut butter on most anything (bread, crackers, apple), carrot or celery sticks (older toddlers).

2. Fruits, juicesicles, milk, cheese are all good snacks.

3. Eat at more-or-less the same time each day. Constant nibbling is not healthy. It is smart to rinse our mouths after eating to keep teeth healthy.

4. Convenience is more important when snacking. Unfortunately, chips and candy are often more convenient than carrot sticks or an apple, so try to have a container of quick healthy snacks available to toddler (and yourself for that matter). Try to include a dairy product (milk or cheese) and at least one vegetable or fruit with each toddler meal.

5. If potato chips, sodas, candy and other "bad" snack foods are in the house we will eat them. Incidentally you don't need to worry about fat for the toddler. A toddler needs fat for proper development, even though you do not.

6. Food is for nutrition and should not be used as a bribe, reward or punishment.

7. Cooking, eating and taking your food to interesting places for picnics can provide opportunities to use real-life experiences to learn wonderful things.

Older children enjoy chopping, measuring, pouring, using ovens and following simple recipes. The fun and learning opportunities are endless.

Toddlers are less interested in food now because their growth has slowed. Each day toddlers need to eat about: 2–3 servings of meat or meat alternatives, 6 or more servings (1/4 or1/2 slice) of breads, cereals and grains, 4–6 servings (1/2 C) of dairy products, and 5 or more servings (2 Tbsp) of fruits and vegetables.

From: *Science, Math & Nutrition for Toddlers: Setting the Stage for Serendipity* by Rita Schrank, Humanics, Atlanta, GA

Fire Safety

Dear Parents of Toddlers:

Now that your toddler is on her feet, she will be more easily "getting into things." Take a look around your house and be sure it is fire safe.

Watch those flames!

- Keep matches and lighters out of a toddler's reach. Toddlers are already able to start flames.
- Praise your child when she brings you matches or lighters which she has found.
- Teach your toddler that fire is dangerous. It is tool, not a toy.
- Explain that grown-ups use matches to light candles, start a campfire or light a barbecue.
- Teach older children how to respect fire, and perhaps even provide opportunities for them to light matches under a grown-up's supervision.

Do you have smoke detectors? When was the last time you checked them?

During the holidays:

Keep open flames away from holiday decorations—especially the Christmas tree. A freshly cut tree needs water every day. The tree is even more of a fire risk when it starts to lose its needles.

Trees, wreaths and gift wrapping should never be burnt in the fireplace or wood stove.

Don't give fire a place to start!

A useful reference book: Gail Gibbons' *Fire! Fire!*, Holiday House Book.

From: *Science, Math & Nutrition for Toddlers: Setting the Stage for Serendipity* by Rita Schrank, Humanics, Atlanta, GA

More Math Fun

Dear Parents of Toddlers:

We cannot *teach* toddlers math. However, we *can* take advantage of daily experiences to prepare toddlers to understand math ideas because toddlers enthusiastically taste and test their new world. Even if you are not "good at math," you can help your toddler learn some math basics.

> *"Toddlers know how small they are and how big grownups are."*

The Notions of Ratio, Scale and Numbering

Because toddlers are seeing the world through their own eyes, with few experiences under their belt, they learn new concepts in relation to themselves. Toddlers' early measuring often uses their own bodies as reference—i.e., "Build it as tall as I am," "It is as big as my finger," "Daddy is bigger than I am," or "I love you this much!" An arm span is a wonderful unit of measure!

We want to establish the notions of ratio and scale early on in math learning. Our society suffers from innumeracy and as a result, we tend to be penny-wise and pound-foolish. Understanding ratios ("How big am I compared to Daddy?") and understanding scale ("How many of me would it take to make one Daddy?") are important basic concepts. Both notions can be illustrated visually—like measuring children & ourselves and regularly making height marks on a wall. Using actual numbers and units of measure can come later.

Make an effort to talk about comparisons—tall/short, few/many, less/more, full/empty.

Books to Help Introduce Toddlers to Math Ideas:

Sara Atherlay's *Math in the Bath. (and other places too)*.

Sam McBratney's *Guess* √ *How Much I Love You*.

Judy Feldman's *Shapes in Nature Eyewitness Juniors*.

A little old for toddlers, but lively, and one of the few books that begin to address simple math concepts for young children.

David and Wendy Clemson's *Mathematics*. NY. Dorling Kindersley.

Colorful. Good book to "talk about."

From: *Science, Math & Nutrition for Toddlers: Setting the Stage for Serendipity* by Rita Schrank, Humanics, Atlanta, GA

Toys

"Play is the best way to be free to wonder, discover and get experience." —Bev Bos

Dear Parents of Toddlers:

Toys are your toddler's tools for learning and they can be expensive, so we suggest you take time to choose them carefully.

Children outgrow their clothes because their bodies are growing. In this same way, children outgrow their toys. Try to choose toys that encourage toddlers to improvise, experiment and wonder as they play and discover. Many household items, even cardboard boxes, make great toys.

Learning

As they play children learn to:

> figure out how things work,
> solve problems,
> pick up new words and ideas,
> build strong muscles,
> learn to cooperate with others.

There are some playthings that last an entire childhood: blocks, sand, water, clay, and musical instruments. Art materials build skills that lead to reading, writing and noticing beauty in life. Even young toddlers can hold a crayon or a watercolor marker and make lovely "pictures" on a large sheet of newsprint.

Pretend toys give children a chance to try new behaviors where they make the rules. They can help children understand the world and how we can work together. Offer some inexpensive real things like adult hats and purses. Look around your house and watch your toddler as he imitates adults, and you will see a wealth of ideas for pretend play. Try to supply toddlers with small safe duplicates of the tools you use—they will enjoy "working" with you.

Getting the Most From Toys

You set the stage for how and what your children will learn with the toys and materials that you select. If you can, arrange toys on low shelves so that toddlers can locate, reach, and return the toys. Learning to return them to where they belong is a part of the play. You may want to keep some toys stored to be brought out later.

Don't Forget Books!

Give your toddlers the love of books and you have given them a gift of a lifetime.

Safety:

Safe Toys for Children are:
• without button eyes or small parts that can come off,
• well made with no splinters or sharp parts,
• shatter-proof,
• easily cleaned,
• kept out of reach of younger children,
• checked frequently for safety.

Share the Chubby Book: *Let's Play; Vamos a Comer* with your children.

From: *Science, Math & Nutrition for Toddlers: Setting the Stage for Serendipity* by Rita Schrank, Humanics, Atlanta, GA

Caring About Our Environment

"Making do" can be fun, believe it or not.

Recycle

√ Let your toddler "help" you recycle paper, plastic, glass, aluminum and batteries. As you put each item into its proper bin you are helping your toddler categorize—a special math skill.

Cut Down on Pesticide Use

Toddlers will be happy to help you get rid of pests "organically." Snail hunts are lots of fun. You might even bring a couple of snails into a simple terrarium for a short visit. Snails are happy living on dirt with a little water. They eat lettuce.

Use Public Transportation

Toddlers love buses.

√ Leave Parks and Campgrounds Better Than You Found Them

Toddlers loves to help pick up litter and put things into trash cans—a good habit to develop.

Enjoy the Beauty Around You

Nature provides us with some wonderful sights. Enjoy them with your toddler. People who enjoy nature seldom wreck it.

Buy Smart, Use Smart—Reuse and Recycle

Watch your toddler use things in ways you'd never expect. Give him an empty paper towel roller, the wrappings from a gift, or any "clean" trash and watch what happens.

Some of the most "beautiful pictures" have been made on the back of a piece of recycled paper. Toddlers think they look great on the refrigerator.

"Make Do"

You'll be surprised at what fun it can be to "make do." See who can be the most inventive in your home. Remember the old saying "Necessity is the mother of invention?"

From: *Science, Math & Nutrition for Toddlers: Setting the Stage for Serendipity* by Rita Schrank, Humanics, Atlanta, GA

Science Activities

"I have no special talent. I am only passionately curious." —*Albert Einstein*

Reader's Objectives

_____
_____
_____
_____
_____
_____
_____
_____
_____
_____
_____
_____
_____
_____
_____
_____
_____
_____
_____
_____
_____
_____
_____
_____
_____

Introduction to the Science Activity Pages

There are at least three ways to use the Activity Pages: 1) go through the pages and make a mental note of the ideas and resources required, catch the spirit, and keep your eyes open for those "teachable moments" as they present themselves; 2) browse through the pages looking for age-appropriate ideas, and set up a project (set the stage) for you and toddler to do; or 3) when you come upon something interesting like, say a bug, look in the INDEX to see what activities are suggested. When you have truly caught the spirit of serendipity* you will no longer need the Activity Pages.

The Scientific Method

Lecturing is a totally ineffective way to teach toddlers. Educators realize that science learning is more effective when observing, comparing, classifying, predicting, and interpreting, and with toddlers, talking about all of it.

Scientific thinking is stimulated as a child tests out ideas: "What will happen if...," discovers and invents new forms, or compares and classifies different sizes and shapes. As a toddler becomes familiar with the properties of the world around, he learns to use clues to predict certain outcomes—this is thinking scientifically.

There is a difference between looking and seeing. Help a toddler see what he is looking at. Give him a label (dog), and describe it ("big dog with long brown hair. He is friendly"). While you are at it, start looking at old familiar things afresh. You will enjoy it.

Rachel Carson, in her *Sense of Wonder*, said, "A child's world is fresh and new and beautiful, full of wonder and excitement. It is our misfortune that for most of us that clear-eyed vision...is dimmed and even lost before we reach adulthood." Lets help our toddlers keep their sense of wonder, and lets try to regain ours.

*Serendipity: Good luck in discovering agreeable things by accident."

—Webster

Reader's Lesson Plans

Listening - P.29
Water - P.75
☆ Counting note on p.87
P.109

☆ Whispering - P.33
Cooking - 66
P.89
P.112

Spiders - P.44
Math - P.81 - Duplo
P.93 - Tape measure
P.117

Let's Go Outside P.53
P.83 - Eating shapes
P.94

Balloons

Things You Will Need:

Balloons

What To Do:

Choose the color of balloon that the toddler wants. Talk about the balloons. Show how they stretch and snap.

Blow up the balloons. Talk about how they get bigger and smaller. Let toddlers feel the air escaping, not that you nor the toddler can see the air.

Ask or tell the toddler before you let the balloon loose to become a "rocket"—little ones can be startled. Balloons that pop are also startling. Be prepared for tears.

Tie the balloons so the "air cannot escape." Bat it, kick it or toss it. Enjoy it and talk about it.

If you have a helium balloon, tie a small figure to the bottom and "give it a ride." Sing "Up and away in my little toy balloon."

New Words:

color	air	escape	
blast off	balloon	tie	kick
blow	rocket	bat	
breathe			

Sample Comments & Questions:

"What color balloon would you like?" "How big shall we make the balloon?" "Can you feel the air come out of the balloon?" "You can feel air but you can't see it." "Shall we put lots of air or just a little air into the balloon?" "Shall we make the balloon go off like a rocket? Are you ready?"

Save for Older Children:

A balloon can be a container of air or water, a surface holding an electric charge and elastic membrane. Older children can blow up their own balloons. Making a square balloon by freezing a water-filled balloon in a box and discussing laws of force and reaction can be saved for older children. Putting soda in a balloon, fixing it to a neck of a bottle with vinegar in it and shaking will cause the balloon to expand.

*Note:

Balloons are the least expensive, most colorful toy available, but if popped can be extremely dangerous for little ones who are still putting things into their mouths. Therefore be sure to supervise and keep track of the broken pieces!

Things That Fly

Things You Will Need:

An airplane in the sky

An airport nearby

Birds or butterflies in the sky or nearby

Toy airplanes and pictures of planes

Useful Resource Book:

Gibbons, Gail, *Flying*

Musical Resource:

Raffi's *Riding in An Airplane*

What To Do:

Keep you eyes out for flying things. Watch, enjoy and talk about them.

Pretend to be an airplane—make those wonderful engine noises.

Visit an airport and try to let toddler sit in an airplane. Watch planes taking off and landing.

New Words:

flying   airplane   real/toy

bird   sky   motor noise

Sample Questions and Comments:

"What do you suppose that is in the sky? Is it a bird or an airplane?" "Sometimes I wish I could fly." "Could we pretend to fly? Can you make a sound like the airplane motor?" "Airplanes make a lot of noise, don't they!" "Can you think of anything else that flies?" "I saw a beetle flying once—a ladybug beetle." "Bees and flies fly too, don't they?"

Save for Older Children:

Talking about kinds of airplanes, blimps and balloons. Talking about laminar forces and the Bernoulli effect will be fascinating later.

Note:

See also Ornithology/birds, Entomology/butterflies & Physics/air.

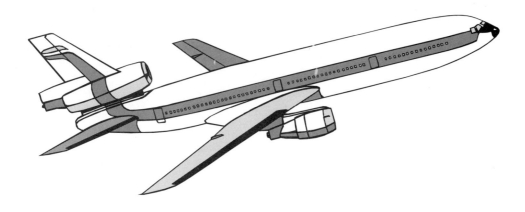

Counting Body Parts

You Will Need:

No materials are needed.

To Do:

Play a counting game. Put your finger on a body part—nose. Say, "Nose. One nose." Put your fingers on your ears. "Two ears. One, two." "Touch your feet. One, two feet."

Books to Share:

Rockwell/Williams/Sherwood's *Everybody Has A Body: Science From Head To Toe*,

Robert James' *Nose* has some nice pictures, Gillmard Jeunesse and Pascale de Bourgoing's *human body* and Edith Baer's *The Wonder of Hands*.

New Words

names for body parts		clap
shoulders	knees	toes

Save for Later

Counting more than two objects other than toddler's body parts. Rote counting and counting rhymes are fun for toddlers. "Counting rationally" comes later.

Tip:

Toddlers may need to watch and then practice, before they can play this game without having you to copy. It is natural to want to go right on to the "Hokey Pokey" after playing this game.

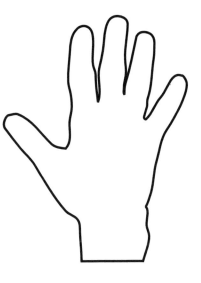

Note:

If humans had come fitted-out with four fingers on each hand, then we would have undoubtedly developed a number system on a base eight rather than base ten. Our decade would be octades. We wouldn't talk about the Gay Nineties, or the Roaring Twenties. Since we have ten fingers and ten toes, the toddler will learn our number system on base ten..."personally relevant."

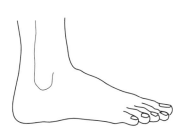

Do You Hear What I Hear?

You Will Need:

A quiet time when toddlers can listen—maybe at snack time.

Good Resource Books:

Aliki's *My Five Senses,* Stage 1. Let's Read and Find Out Science Series

Snapshot™ Books. *Super Senses*

Bill Martin's *Listen to the Rain*

To Do:

Listen to see what you can hear—the crunch of an apple as we bite into it, rain on the roof, birds singing, a train, children talking at the next table.

New Words:

Descriptions of sounds

Apples crunch.

Birds tweet or chirp.

Rain sounds like fingers tapping on the roof, etc.

Children giggle, sing, shout and whisper.

Sample Questions and Comments:

"Let's listen to see what we can hear." "What do you hear?" (whisper) "I hear the children talking at the next table." "Do I hear a bird?" "I just heard a bell—It must be time for the students to change classes." "Close your eyes. Guess what is this sound?" (ring a bell, clap your hands, whistle). "Can you guess what this is?"

Save for Older Children:

Talking about sophisticated sounds like a fan whirring.

Note:

See also Physics/naming sounds, Anatomy/senses/listening.

Footprints

You will need:

Bare feet

Water

A place to make footprints (cement, asphalt)

Resource Books

Aliki's *My Feet, My Hands* (Let's Read and Find Out Science Book)

Angela Johnson's *Rain Feet*

Peter Parnall's *Feet!*

What To Do:

Make water footprints on the pavement.

Talk about it (wouldn't it be nice if a dog or cat walked by?)

Distribute paintbrushes and "paint" everything in sight with water.

Have you ever tried to make a "footprint" with the bottom of your fist, adding toes with fingerprints?

To extend this activity further, you can talk about animal feet: *"Feet of snails are only one. Birds grow two to hop and run. Dogs and cats and cows have four. Ants and beetles add two more. Spiders run around on eight, which may seem a lot, but wait…centipedes have more than thirty feet to wash when they get dirty."* —Aileen Fisher

New Words:

foot/feet	footprint	paintbrush
pavement/sidewalk/cement		water
bare feet	walk/step	bigger/smaller
evaporate		

Sample Questions or Comments:

"Look at your footprint." "Whose print is that?" "Whose footprint is smaller?" "Whose footprint is bigger?" "Where did my footprint go? It evaporated into the air. Let's do that again." Clapping game: "Feet, feet, down the street."

Note:

This activity can be extended by using washable paint on feet, stepping on paper—messy, but a favorite of grandparents. Paper towels in the tray before adding the paint will make a "stamp pad" into which toddlers can coat their feet with paint. Toddlers love to clean up after this project.

Go Go Go

You will need:

No materials are needed—an open space with room to move (a lawn or carpet).

To Do:

Walk. Run. Crawl. Walk backward. Jump. "Race,"* Enjoy it and talk about it.

If you see a toddler running, start running alongside her. Talk about it. Change the pace—run slowly, run fast, walk...play Stop & Go.†

Older toddlers may like to try walking backwards.

Draw a chalk line and show how to walk toe to heel.

Resource Books

Usborne Books at Home, board books for babies and toddlers: *My Body*

Aliki's *My Feet, My Hands* (Let's Read and Find Out Science Book)

Audrey Wood's *Quick as a Cricket*

New Words:

run	walk	stop/go
crawl	toe and heel	slow/fast
together		race

Note:

A toddler is happiest when she is moving. The human body is a wonderful thing.

*Save for Older Children:

Competitive races that have winners and losers.

†Later the game of "Mother, May I?" will be fun to play.

Listening

You will need:

Vacuum cleaner hose

Empty paper towel cardboard tube

Toilet tissue tubes for each child (optional)

and/or

wrist and/or ankle bells

music source

Resource Books

Usborne Books at Home: *Who's Making That Noise?* Board books for babies and toddlers

Helen Lester's *Listen Buddy*

Bill Martin's *Listen to the Rain*

To Do:

Talk into the hose and put the hose to the toddler's ear and let her listen.

Talk, whisper or hum into the cardboard tube held close to her ear. Give the toddler a tube and let her do the same. Place both hands over your ears and ask her to do the same. What is the reaction? Does she understand that her ears are doing the hearing?

—and/or—

Show the toddler how to shake the wrist bells by herself and then shake the bells to music. If you are lucky, you will have enough to put some bells around the ankles. March music is especially fun with bells around your wrist and/or ankles. Talk about the noises and music you are making.

New Words:

hose   tube   noise   wrist (ankle) bells

music   march

Sample Questions or Comments:

"Hello, John." "Can you hear me talking through the hose?" "The noise goes right through the tube." "Can you hum into the tube so that I can hear you?" "Listen to the bells when we shake them." "Let's march to the music. Shake those bells!" "What lovely noises we are making!"

Save for Older Children:

String telephones, funnel telephones and other enjoyable sound experiences. Toddlers can join preschool "orchestras" and play a number of shakers, bells, etc. They do best when they have models to follow.

Note:

If you want children to listen to you, start by listening to them…"with your eyes," as my three-year-old son once said to me.

Naming Body Parts

You will need:

No materials are needed at first
A doll or stuffed animal

Resource Books

Rockwell/Williams/Sherwoods's *Everybody Has a Body: Science From Head to Toe*

Board books for babies and toddlers: Usborne Books at Home, *My Body*

Gillimard Jeunesse and Pascale de Bourgoing's *human body*

Edith Baer's *The Wonder of Hands*

Music Resources

Marino & Houlihan's *Mother Goose Time: Programs for Babies and Their Caregivers,* including such favorites as "Head, Shoulders, Knees and Toes," "Clap, Clap, Clap Your Hands."

Tom Glazer's *More Music for 1's and 2's,* includes "Where Are Your Eyes?" and "What Does Baby Hear?"

To Do:

Play "Where is your (body part)" game.

Sample Comments and Questions:

"This is my nose. Where is your nose?" "Where is the doll's (puppy's, etc.) nose?" "Can you show me your eyes?" "Where is your mouth?" "What do you do with your mouth (eat, chew, smile)?" "Where are your hands? Hold one hand up. Hold the other hand up. Wiggle your hands (show them). Shake your hands." OK!

"This little piggy went to market. This little piggy stayed home. This little piggy had roast beef. This little piggy had none. This little piggy went, 'Wee, wee, wee!' all the way home."

New Words:

names for body parts
smell, sniff
doll & stuffed animal's labels & names
wiggle toes
eat, chew, smile

Note:

This is an opportunity for intimacy and quiet. Talk about one feature at a time and be sure to touch the feature as you name it. This is a good game to play during diaper changing time.

Save For Older Children:

Toddlers love watching finger plays and hand games and will play along if they are not in the spotlight. Activities where unusual parts are identified, e.g. stomach, chest are interesting for older children. Playing the "Knee chest" song-game (toddlers like to watch this game too). Using a mirror to identify body parts—the mirror can confuse a young toddler. Talking about masticating and that enzymes are involved in digesting food.

Smells

You will need:

Anything that has an obvious smell, e.g. snack cooking, onions, smelly socks, perfume, vanilla (high alcohol content—be careful that toddler does not try to drink it), spoiled food (old cabbage smells pretty bad)

Cotton balls

Milk for vanilla milk (recipe—one wee drop per cup), spoon to stir and cup

To Do

Take note of the smell. Liquids can be poured into a cotton ball for easier smelling. Talk about it. If you like, make vanilla milk for snack. Sugar enhances the taste of vanilla milk, but it is not necessary. It makes a good warm drink just before nap.

New Words

smell/smelly	spoiled	ugh
delicious	nose	vanilla
onion	cotton ball	

Sample Questions or Comments:

"What is that smell? Let's find out what it is." "Ugh, that fish is spoiled. It smells bad." "This is vanilla. We use it to make cookies and cakes taste good. At snack time we can have vanilla milk if you like." "Mmm, something smells good. It it the snack cooking?"

Save for Older Children

Wait on letting toddlers smell ammonia and moth balls. They have such penetrating smells it is tempting to impress toddler with them, but both are a bit too strong and toddlers might try to eat the moth balls.

Smells like herbs, bath foam, pine cones, orange peel are too subtle for toddlers. You will notice that when asked to smell a toddler will pretend to smell but will not automatically draw a breath. Just as we learn to blow out as we blow a candle, we also need to learn to breathe in as we smell. Talking about staying downwind of a wild animal if you don't want to be noticed is interesting to a much older child, as are competitive smell guessing games.

Good Resource Books:

Snapshot™ Books, *Super Senses,* (has colorful photographs)

Robert James' *Nose*

Usborne Books at Home, *Who's Making That Smell?*

Kathie Billingslea Smith and Victoria Crenson's *Smelling*

Aliki's *My Five Senses,* Stage 1. *Let's Read and Find Out* series

Note:

Smell and taste are closely associated. Often when we think we taste something, we have really smelled it. We can only taste sweet, salt, sour and bitter. On the other hand, our noses can pick up thousands of separate smells. See also: Nutrition/Tasting new foods and Anatomy/senses/tasting.

Toileting

You Will Need:

A bathroom or a potty
Diapering time

To Do:

You may not find toileting a fascinating activity, but toddlers will be interested—for now as a new "scientific" experience or later with the prospect of escaping diapers.

Give toddlers an opportunity to watch others using the toilet and talk about what is happening.

Listen for the tinkle, hear the plop, flush the toilet, note the loud noise, wash hands (lovely water!).

Talk about the physical sensations.

During diapering time talk about what happened, "You made a big poop." "Wet diapers are cold." "The urine leaked around the diaper and came running down your leg." "Poop is heavy to carry around in a diaper." Also talk about toilets and how they work and how useful they are.

Books:

Books about toilet learning abound in the bookstores and libraries, including:

Taro Gome's *Everyone Poops*

Rockwell/Williams/Sherwood's *Everybody Has a Body: Science From Head to Toe*

Music:

"Today I Took My Diapers Off," from Hap Palmer's *Babysong*. Cassette/guide and compact disc/guide.

Signing time 's "Diaper Dance"

New Words:

function names	toilet/potty
wash/soap/water/clean	flush
brown/pooh/B.M.	yellow/urine
noise	

Tip:

Toddlers need to know that they are in command of this bodily function. When they succeed it will be at their own pace and on their own terms. They gain control of the bladder and sphincter muscles between 1 1/2 and 2 1/2 years of age, bowels before bladder, daytime before nighttime.

Note:

This is an opportunity to use the scientific method to learn a useful skill and to talk about our bodies, "What do you suppose happens when we...."

Whispering

You Will Need:

Your own voice and toddler's ears.

Resource Book:

Helen Lester's *Listen Buddy*

To Do:

Whisper (a stage whisper at first), so everyone can see you.

Whisper into a toddler's ear.

Talk about small voices and big voices (Everyone talks about inside voices and out-side voices.).

You can extend this activity by adding music and singing softly and loudly.

New Words:

whisper   shout   voice
listen/hear   ears

Sample Questions and Comments:

"Can you hear me whisper?" "Can you whisper?" "When you whisper you have a very small voice." "You sound like the wind." "Can you speak in a very big voice?" "Can you shout?"

Note:

If you want someone to listen, whisper or speak in a very low voice.

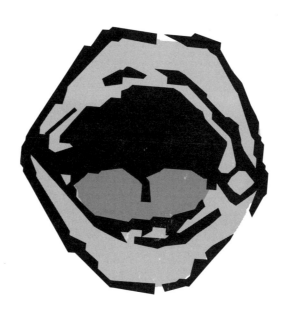

Yum!

You Will Need:

Things to taste

Good Resource Books:

Aliki's *My Five Senses,* Stage 1. Let's Read and Find Out Science Series

Snapshot™ Books, *Super Senses* (a colorful boardbook of photographs)

Smith, Kathie Billingslea, and Victor Crenson, *Tasting*

To Do:

Taste foods. It is best to taste one new food at a time. Wait until the child is hungry before trying new foods. Talk about how food tastes, what it feels like, and how it smells. This will help to keep you from saying, "Eat your food. It is good for you!"

New Words:

Descriptions of foods:

Apples are juicy, tart, crisp.
Ice cream is cold, smooth, sweet.

Crackers are salty, dry, crunchy.

Artichokes are green. We dip the leaf in butter. We make teeth marks when we scrape the good part off the leaf.

Sample Questions and Comments:

"This apple is crisp." "The cracker is salty." "Doesn't milk taste good after a peanut butter sandwich?" "I like sugar on my strawberry. It is sweet."

Save for Older Children:

Talking about how your tongue tastes the salt and sends messages about the texture of the food, and the nose smells the aroma and tells the brain…and all that "scientific" stuff!

See also: Nutrition/new foods

Note:

A baby can taste foods very well. They have taste buds on their tongue, inside their cheeks, and on the roof of their mouth. As we grow older, the taste buds on the sides and roof of the mouth disappear. The ones on the tongue remain.

Diggin'

You Will Need:

A place to dig (If you are lucky you might find a place where old shells can be found, but any digging place will do for a toddler).

Shovel, trowels, archaeologist's hammer (for adult), old brushes, collecting bag, an "archaeologist's" framed screen or colander to sift dirt.

—or—

A sand pile where you can hide toy dinosaurs, shells or other treasures.*

Resource Book:

Paul D. Taylor's *Fossil,* another beautiful book of eyecatching photographs by Eyewitness Books

To Do:

Look for treasures in the dirt or sand. You will be surprised what you can find; even rocks can be interesting. Pour a shovelful of dirt into the colander or screen and sift. The dirt will sift through and interesting things will stay in the colander. Take care as you unearth "the treasures" to keep them as whole as possible. Toddlers will of course have no respect for real treasures, so be aware. Talk about looking for a place to find old fossils or artifacts. You may want to "plant" a few items just to insure success—even plastic

dinosaurs will do. You can add water if you are ready for a mess. Water unearths different items. Don't be afraid of the real words—toddlers are mimics and will have fun with words like "digs," "dinosaurs," "shell."

New Words:

digging	screen	digs
colander	treasure	brush
artifact	sift	shell
mud		

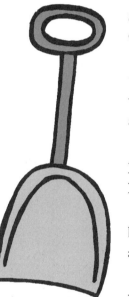

Sample Questions or Comments:

"Let's go look for treasures." "Look what Jake found." "How did you find that Jake?" "What do you suppose this is?" "Let's find out."

Neat Things to Save Until Later:

Visiting a real "dig," reading books, watching tv and talking about ancient civilizations.

Note:

The young toddler will not understand much of what you are saying and doing, but will enjoy the digging and will eventually put it all together.

"Fossils"

You Will Need:

Soft soil colored playdough*

Shells

Plastic knife (supervise)

or

Toy dinosaurs, chicken bones, other possible "fossils"

Useful Books:

Paul D. Taylor's *Fossil,* Eyewitness Book with beautiful photographs

Board Books for Babies and Toddlers: *Playdough*

Aliki's *Digging up Dinosaurs*

New Words:

fossil   shell   knife

cut

paleontologist

find/discover

To Do:

Press shells into the dough and cover. Then go "looking" for the shells. Talk about marks that the shells make in the playdough. Cut into dough and look at a cross section.

Objects can be hidden in the the sand table and

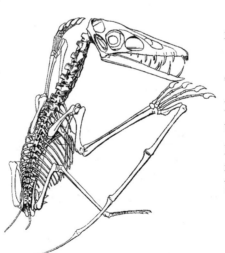

"found" using an "archaeologist's" screen to sift out the treasures. Pairs of objects can be hidden, and older toddlers can try to unearth matching ones.

Tip: If the dough is soft enough you can use leaves and other natural items to hide and discover.

Sample Questions and Comments:

"I wonder what we will find inside this playdough?" "What do you suppose that is?" "Is that a shell?" "Let's look at our shell book to see what kind of a shell this is." "Could this be a fossil? It looks very old."

Interesting Things to Save for Older Children:

Real fossil hunting

Identifying and categorizing real fossils.

Reading books about fossils and artifacts.

*Note:

It is best to save playdough until the toddler has stopped putting everything into her mouth. Then playdough will be a constant favorite.

Star Gazin'

"Those who dwell among the beauties and mysteries of the earth
are never alone or weary of life."

You Will Need:

A clear dark night with the moon and stars
are in sight.

Resource Book:

Anna Hines' *Sky All Around**

To Do:

Take warmly dressed toddler in arms or by
the hand, and go out into the night. Look at
the stars and catch that intimate moment to
look and listen. It is a good time to whisper.

Read: Martin Waddell's *Can't You Sleep Little
Bear?*

Rechel Carson's *Sense of Wonder*

New Words

Dark	Moon/Full Moon
Stars	Quiet
Wind in the Trees	Far away

Sample Questions or Comments:

"It is dark tonight. This is a good night to
go star gazing." "Look at all the stars." "The
stars are very far away." "The moon is almost
full tonight." "The moon is full tonight. Let's
go look." "What do you hear?" "I think I hear
wind in the trees." "I hear some crickets
singing."

Save For Older Children:

Talking of planets, light years, comets and
planetariums. Finding the moon during the
daytime.

*Anna Grossnickle Hines' *Sky All Around* is a
delightful story designed to introduce
younger readers to the joy of discovering
objects in the night sky.

Note:

Telescopes or binoculars: a toddler will
enjoy watching you look through them, but
has not developed the skill of seeing through
a telescope or binocular yet. Some enjoy pretending to look through a toilet paper tube
"telescope." A telescoping pretend "spy-glass"
can be made with two cardboard tubes
(paper towel and gift wrap) of different diam-
eters.

Reptiles

You Need:

A small adoptable animal: lizard, snake

Enlarged photos of small animals*

To Do:

Go looking for and adopt a small animal.

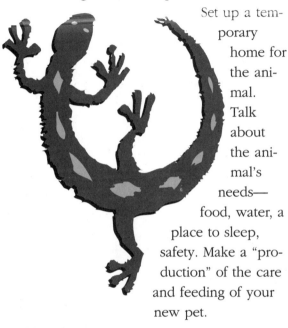

Set up a temporary home for the animal. Talk about the animal's needs— food, water, a place to sleep, safety. Make a "production" of the care and feeding of your new pet.

After the animal has visited for a while, return it to where you found it.

Look up pictures and information about the animal. Share Pallotta's *The Yucky Reptile Alphabet Book* and Christopher Santoro's *Lift a Rock Find a Bug* with the children. Talk about it.

—*or*—

Care for a pretend creature.

New Words

lizard/newt/frog		snake
rock	soil	water/food
shelter		

Save for Older Children

Talking about and identifying specific species of animals.

See also: Biology/small creatures, Entomology/insects and Invertebrates/worms.

Note:

Protect the small animal from toddler's caresses. A toddler does not know how fragile a small animal can be.

If you cannot have the mess and bother of small animals, you can try to visit a natural history museum where these creatures and their habitats can be seen. Of course they can also be seen in pet shops and zoos.

*The Eyewitness Books and Eyewitness Juniors are rich with large colorful pictures.

Introducing Gardening

You Need:

A small plot of land or a pot of soil

An iris stock, a cutting of jade plant or any other "instant" plant

Water, gardening tools—e.g. trowels, small rake

Resource Book:

Snapshot™ Books, *Things That Grow*

Series of colorful photographic board books

To Do:

Allow plenty of time for digging, breaking clods and working the soil.

Plant (or transplant) a bulb or growing thing like the jade plant or an iris that will produce instant results in their own pot. Then as the toddler tends the plant she will begin to understand that it is a living thing.

A toddler will be delighted to water the plants—in fact drown them, if not supervised.

Getting toddlers truly involved in gardening should start with harvest. They will love to help pull potatoes, radishes, carrots, etc. Then when they are older you can reminisce about the harvest last year, and *then* plant seeds and seedlings.

Gardening Music:

Cathy Fink and Marcy Marxer, *Blanket Full of Dreams,* CD/cassette, award-winning and soothing word pictures like "Sleep like a little pea/Growing in the garden."

Raffi in Concert, includes such toe-tapping songs as "Everything Grows" (Lyrics enclosed).

Save For Older Children:

The wonderful connection children feel for the plants they grow and the animals they meet in their garden can become an everlasting joy as they enjoy successes.

Note:

The toddler will be more interested in cultivating the soil than sowing a thing called a "seed." She will be interested in pouring or running her hand through the seeds and watching you or older children plant the seeds. Younger toddlers have also been known to eat a seed or two.

Recycling

Nobody in this world outshines early childhood people in their efforts to recycle.

You Will Need:

Things to be recycled and a place to sort and hold them

Respect for the earth's resources

Resource Books:

50 Simple Things Kids Can Do to Recycle. Berkeley, CA: Earthworks Press 1994

Jen Green's *Why Throw it Away? Making Mad Machines*

To Do:

Let the toddlers help you clean and sort things to go into the recycling bins. Talk about categories (glass, aluminum, newspapers), and enjoy the process. Include toddlers in your ecological efforts as you buy and as you recycle—talk about using resources wisely and how we can reuse articles.

Let toddlers participate in making simple papier mache from old newspapers, a lovely and messy project.

New Words

clean   glass/green/clear   save
reuse/recycle aluminum   newspaper

Sample Questions and Comments:

"How can we use this....again? No use wasting it." "Let's put these newspapers in the recycle bin they can be used again." "Let's buy this [product]—they have made it with recycled materials." "Turn off the water for now. We don't want to waste precious water."

Tip:

Clean recycled items make wonderful toys for toddlers—they will use things in entirely different ways than you expect.

Neat Things to Save For Later:

Visiting a recycling plant, turning in recyclables for money.

Dorling Kindersley's *My First Green Book; A Life-Size Guide to Caring for our Environment.* NY: DK Knopf, has wonderful photographs and some good ideas.

Butterflies & Caterpillars

about what it looks like, how many legs it has, and that someday it will be a butterfly or a moth.† After you have had a chance to observe it, let it loose where you found it. Some schools enjoy raising silkworms, but a toddler is not ready to understand what is happening.

You Need:

A bug jar: unbreakable, see-through with a breathable lid

Small paint brush

Enlarged photos of butterflies, moths and caterpillars*

Resource Books:

David Saintsing's *The World of Butterflies* (Where Animals Live Series)

Bobbie Kalman's *Butterflies and Moths*

Susanne Santoro Whayne's *Butterflies*

Eric Carle's *The Very Hungry Caterpillar*

To Do:

Keep your eyes open for butterflies or moths—follow them. Talk about how they are enjoying the flowers. Smell the flowers; how do they smell? Look at pictures of butterflies and moths.

Find a caterpillar and take it home. Use the paintbrush to move or lift a small caterpillar to avoid squashing it. Place it on a piece of paper for easier observation. Talk

New Words:

| butterfly | moth | caterpillar |
| fly | wings/feelers | caterpillar legs |

Tip:

A caterpillar is the larva of either a moth or a butterfly.

Save for Older Children:

Talking about the differences, and identifying species of butterflies and moths.

*Butterfly hatcheries will be more interesting, as well as more fun when the children are older.

For generations preschoolers have had fun making two caterpillars out of an egg carton.

†Note:

Toddlers probably will not understand that butterflies and caterpillars are related but will enjoy watching or chasing butterflies and looking at the caterpillars. However, it may be a challenge to keep the young toddler from eating or squashing the caterpillar. Be sure to wash hands when you get back to home base.

Insects

You Need:

A bug jar: unbreakable, see-through with a breathable lid

Small paint brush

Enlarged photos of insects*

Useful Books:

Barrie Watts' *Ants; Beetles;* or *Ladybugs*

Bobbie Kalman Books, *Bugs & Other Insects*

Millicent Selsam's *Backyard Insects,* big beautiful photos and text full of information.

Christopher Santoro's *Lift a Rock Find a Bug*

To Do:

Go looking for small crawling creatures (under a rock, on a bush, in the grass, in an anthill)—sowbugs and ants are the most available "bugs." There are over 250,000 different kinds of beetles alone, including the ladybug beetle.

Find an anthill or some ants eating a carcass. Watch the ants take the food to their household and talk about it.

Take home a specimen or two. Use the paintbrush to move or lift small insects to avoid squashing them. Let harmless insects loose where you found them after you have had a chance to observe them.

New Words

bug   insect   rock   soil

name of insect (beetle, ant, earwig)

hiding/eating/hopping

Tip:

Any bug that has more than three pairs of legs isn't an insect. It could be a spider or centipede.

Save for Older Children:

Talking about and identifying species of insects.

Discussing the difference between spiders and insects.

Ant farms and butterfly hatcheries are lost on toddlers and will be interesting and more fun later.

Talking about harmful bugs (mosquitoes, bees, fleas, ticks) can be saved for later also.

Note:

Toddlers may not understand what is happening yet, but will enjoy the process. However, toddlers can see and pick up the tiniest of things. It will be a challenge to keep young toddlers from eating the bugs. Be sure to wash hands when you get back to home base.

Eyewitness Books are rich with large colorful pictures.

Small Creatures

You Need:

A large bug jar: unbreakable, see-through with a breathable lid. Enlarged photos of small animals.* A small adoptable animal, such as a bug, snail.

A large clear container for a temporary home (fish bowl or aquarium); rocks, grass or plants; a small amount of water.

Resource Books:

Henwood, Chris, *Snails and Slugs* (Keeping Minibeasts Series)

Jennifer Coldrey's *Discovering Slugs and Snails* (for help with terrarium)

To Do:

Go looking for a small crawling creature.

Take home a specimen or two. Create a temporary home in the clear container for the animal. Talk about the animal's needs: food, water, a place to sleep, safety. Start with a little gravel on the bottom, then an inch of soil, add rocks, leaves, grass or plant. Add water to dampen.

Look up pictures and information about the animal. Talk about it.

Read Christopher Santoro's *Lift a Rock Find a Bug*.

After the animal has visited for a while return it where you found it.

New Words

snail/slug	bug	rock	soil
water/food	shelter		creature

Save for Older Children

Talking about and identifying specific species of animals. Frogs/tadpoles/polywogs are a whole new adventure in themselves.

Note:

The toddler may not understand what is happening yet, but will enjoy hunting for and watching you bring home the specimen. Try to protect the creature from mutilation—a toddler cannot be careful yet. Be sure to wash hands when you get back to home base.

If you cannot have the mess of small animals, you can try to visit a natural history museum where little animals and their habitats can be seen.

*The Eyewitness Books and Eyewitness Juniors are rich with large colorful pictures.

Spiders

You Need:

Foggy dewy day

A place where spiders live (especially golden orbs)

Resource Books:

Carle's *Very Busy Spider* (Toddler can feel the raised webs in the book).

I can Read About *Spiders*

Dr. David Darling's *Spiderwebs to Skyscrapers: The Science of Structures*—helpful for the adult who gets questions about how things are built

Barrie Watt's *Spider's Web*

To Do:

Go looking for spider webs. If possible, quietly get behind the toddler, take his hands in yours (He will want to grab the web) and quietly talk about what you are seeing. If you are lucky the spider will be nearby.

Because children are fascinated by spiders, preschools build whole curriculum around them—building webs of rope, pretending to be spiders, etc.

Sing the old standby "Eensy Weensy Spider."

New Words:

web   spider   dew/water
fog/mist/wet   sparkle home

Sample Comments and Questions:

"Let's see if we can find a spider's web." "Look at the spider's web." "What a beautiful home the spider has made!" "There she is!" "Look at the water drops sparkling in the sun!"

Lots of Things to Save for Older Children

Looking for insects that are wrapped in the web and explaining that the spider is saving it for a future meal—we save our food in a refrigerator or freezer.

Weaving, spiders, halloween—spiders are a favorite of preschoolers. Talking of poisonous spiders will be interesting and important for older kids (especially 4 year olds). Older children love White's *Charlotte's Web.*

An interesting project to watch a teacher do—spray the web with white paint and take a "print" of the web on black construction paper.

It is interesting to pound golf tees into styrofoam and string a "web" around them.

Worms

You Will Need:

A corn cob after the kernels have been eaten

Moist soil in which to bury the cob outside

A fish bowl or old fish tank

A few pebbles or gravel, soil, small green plant, water, black construction paper, scotch tape and lettuce, other old green veggie, or a pinch of cornmeal

Resource Books:

Dorling Kindersley's, Ltd. *My First Look at Nature*

Chris Henwood's *Earthworms*

Chris Henwood's *Snails and Slugs*, (Keeping Minibeasts Series)

To Do:

Bury the corn cob. After a few days "discover" it. You will find worms. Bring home some worms and place them into the fish bowl that has been filled with moist soil with a few pebbles on the bottom. Plant the greenery in the soil and add a bit of water. Add worms. Tape black paper over one side of the bowl. Later, when you lift the paper, you will find the worms have made a home.* Sow bugs, snails or other creepy crawlies that a toddler might bring to you can go into this cozy

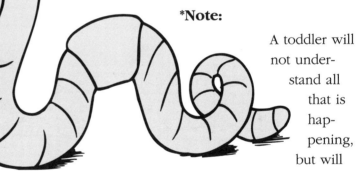

home. Keep this environment moist. Feed leaves or tiny amounts of decaying veggies to the worms.

New Words:

| worm | sowbug | soil/dirt |
| water | cornmeal | |

Tip:

Worms live on land and in the ocean, ponds or streams. There are about 7000 varieties including roundworms, flatworms, hair worms, ringed or segmented worms. The reason that worms come up on top of the soil on rainy days is because they can't breathe in water soaked soil.

*Note:

A toddler will not understand all that is happening, but will enjoy the burying, and the subsequent worm hunt; bringing the worms home and pouring them into the fish bowl. Toddlers also love to simply dig in the dirt or enjoy helping gardeners. Gardeners be aware that a toddler will enjoy plucking small seedlings from the garden if not supervised. Be sure to wash hands often.

Accidents Will Happen

You Will Need:

A small crisis—a bug bite or a scraped knee

Resource Books:

Souza, D.M. *What Bit Me? Creatures Around Us*

There are many books to help talk about seeing a doctor including Margot Linn's *A Trip to The Doctor.*

To Do:

Keep calm. Take the opportunity to talk about what we do in emergencies. Include any watchers in the experience. Talk about how doctors, nurses, teachers and mothers help us when we are hurt. Use real words—but not too many.

Talk about what you are doing. Talk about cleaning a wound and keeping germs from making the sore infected. When you have a chance, talk through parts of Judith Rice's *Those Mean Nasty Dirty Downright Disgusting but...INVISIBLE Germs.* Use the book to talk about washing hands (before cooking, before snack, etc) because of germs.

New Words:

ouch	wound	sore
clean	bleeding	medicine
bandaid	pain	ice/cold
germs		

Sample Questions or Comments:

"What happened?" "Lets go inside and…" "The ice is cold. It will help take away the pain."* "We will take you to the doctor. She will make you better."

*Note:

The ice reduces the nerve sensitivity as it is reducing the swelling.

Save:

Talking about ambulances, broken bones, hospitals and x-ray machines.

Learning to dial 911—an important skill.

Talking about scabs and how they develop and why.

Making a "Get Better" card.

Note:

For now the young toddler will understand only the context of what is happening.

Ouch!

You Will Need:

A wound ("boo-boo" or "owie") such as a scraped knee.

To Do:

Take the opportunity to talk about how the child got "hurt, cut, scraped, bumped..." and include any watchers in the experience. Talk about how grownups—teachers and parents help us when we are hurt. Discuss how to avoid the problem next time. Use real words—but not too many. See also Medicine/Emergencies.

Talk about cleaning a wound and keeping germs from making the sore infected.

Reference Books:

Judith Rice's *Those Mean Nasty Dirty Downright Disgusting But...Invisible Germs*

Harriet Ziefert's *Sam's BooBoo*—comes with a "boo-boo bug" icepack

Melvin Berger's *Germs Make Me Sick.* Also available in a Read-Along book/cassette version

New Words:

ouch	bandaid	pain/hurt
wound	black and blue mark	
sore	bump	infected
clean	germs	medicine
heal	stitches	

Sample Questions or Comments:

"What happened?" "Lets go inside and...." "Next time I'll bet you won't touch the hot stove." "We will clean the wound and put a bandaid on it to keep the germs from getting into it. Then the wound will heal better." "That bump hurt. You will probably have a black and blue mark tomorrow." "Let's see what happens to the hurt tomorrow." "You won't need stitches; we'll use two bandaids to pull the cut together."

Save for Older Children:

Talking about "blood and guts." Your four year olds will discuss them with relish.

First Aid for Wounds:

A rip or tear that bleeds slightly can be treated more casually than a scrape or abrasion or worse, a laceration. A deep, gaping, bleeding cut that damages tissue below the skin should be pressed with a clean cloth at least five minutes to stop bleeding, cleansed with water and antibacterial soap. If the wound is longer than 1/2 inch on the body or 1/4 inch on the face, get to a doctor. Otherwise, apply ointment and cover with a butterfly bandage. Cleanse, apply ointment and put on a new bandage twice daily for two days, and more often if the toddler gets the area dirty.

A Toddler Museum

You Need:

1. An eye for "treasures"—a dead bug, a pretty rock, a shell…The toddler will be the one to find them. Catch her before she eats the treasure!

2. A space to exhibit treasures that the toddler cannot reach without help.*

3. Labels and a pen.

Fun to Have:

Millicent Selsam and Joyce Hunt's *A First Look at Rocks*

Drieger's *Too Many Stones*

Millicent Selsam's *Backyard Insects*

Karen Dawe's *The Pond Book*

To Do:

This "museum" can be a part of your preschool science corner.

Keep your eyes open (toddlers will too) for specimens. This will be the place to keep all those "treasures" that children bring to us with "Look at what I found!" or "What is this?"

Admire, identify, take the specimen to the "museum," label and revisit occasionally. Include watchers in the process. If you don't know what to call the specimen, look it up in your trusty natural history book. Then ostentatiously label it: "Fran's pretty rock" or "Fran's obsidian rock." The museum can grow in specimens and in sophistication.

This method can save the lives of many potential specimens because you can say, "We already have that [creature] in our museum."

Sample Questions or Comments:

"Look what Fran has found. What do you suppose it is?" "Lets find out." "We'll put it into our museum." "At snack time we'll read the book about these bugs." "We already have a [creature] in our museum. We'll let this one go home to his family."

New Words:

specimen	critter names	museum
treasure	label	

Save For Older Children:

*Young toddlers will eat, crunch, pour or sweep everything to the floor of your museum if not supervised. Older children will enjoy having the "science" things available to manipulate. Dead animals in formaldehyde and more sophisticated labeling systems can come later too.

Note:

We have found that toddlers like a natural history museum better than a zoo. The mounted animals stand quietly so that the toddler can see them. Live animals are often sleeping in the corner of their cages.

Natural History Museums

You Will Need:

Time and resources to visit a natural history museum. They are usually child-friendly places. Take water and cups and/or a picnic. Also you should have available picture books and large pictures of the animals you see.

Resource Books:

Eyewitness Books

Dorling Kindersley's *My First Look at Nature*

The Usborne Complete *First Book of Nature*

Wee Pudgy Board Books *Wee Wonders of Nature*

What To Do:

Visit a natural history museum. Zoo animals tend to hide, but museum mounted animals are easier for a toddler to see. Let toddlers lead you. Stop for a rest and a drink. Make the first visit a short one. Return often (with families too) with a picnic. Enjoy and talk about everything you see.

New Words:

animal   museum
names of the animals
mounted/stuffed animals
diorama

Sample Questions & Comments:

"I wonder what that animal is? The sign says..." "Look where the tiger lives." "This diorama shows the wolf's home—it looks like a cozy hole." "I didn't know that bears got so big! This marmot is so little." "Yes, these animals are dead. They have been stuffed."

Save for Older Children:

Formally identifying and categorizing animals

Long explanations about the life and times of any animal

Identifying and talking about animal and bird homes—fun later

Discussing endangered species

Note:

Natural History Museums will offer all kinds of classes and programs to children of various ages. Get on their mailing list.

Air

You Will Need:

Any large piece of tightly woven fabric—parachute, tarp, barbeque cover

Bubble wrap—the large size is best, but small bubbles pop as well

To Do:

Fabric: Float the fabric to the ground while trapping some air (not toddlers) under it. Let toddlers stomp on the air bubbles as the air travels from place to place under the fabric.

Bubble Wrap: Show toddlers how bubble wrap pops when the air is released. Young toddlers will enjoy popping the bubble wrap by driving a trike over the wrap, or putting a block on the bubble and stepping on it.

TALK ABOUT ALL OF IT!

Useful Resource Books:

Walter Kraul's *Earth Water Fire and Air: Playful Explorations in the Four Elements*

Angela Webb's *Air* (A Talkabout Book)

Aliki's *Feel the Wind*

Gail Gibbons' *Catch the Wind*

Sample Questions or Comments:

"The air is trapped under the tarp." "I wonder what will happen if you stomp on the bubble." "Jump on the air bubble and see what happens. The air flows to another place." "See, there it goes—Now it is here." "Look, all the air has escaped." "I wonder what will happen when we help the air escape from the bubble wrap bubbles?" "When the air escapes it pops, when the air escaped from the parachute is just poufed." "Pouf, pop." "Air is funny—you can feel it, but you can't see it."

New Words:

air	trapped	air bubble
bubble wrap		parachute
pouf		tarp
pop	bubble	escaped
air trap		

Save for Older Children:

Kites, blowing out a candle, wind experiments, or talking about air currents and weather.

Note:

As you float the fabric to the ground you will find toddlers immediately ready to stomp on the air bubbles. Very little persuasion will be necessary.

Did you know that air is like water in many ways? It flows, it has currents, and we (and trees too) sway with wind just as water plants do.

Cold Outside/Warm Inside

You Will Need:

Low windows in a warm room on a cold day.

To Do:

Discover steamed windows or blow to make "fog" on the window and talk about it. Feel the temperature of the window and let toddlers draw on the mist.

If it is cold enough you may have ice inside. Scrape it with your nail and talk about it.

New Words:

fog/mist   breathe   cold
outside   inside

Tip:

This can be a good activity when bored riding in a car on a cold day.

Sample Questions and Comments:

"Look at the fog on the window. That means it is colder outside than it is in here."

"Look!—you can draw on the mist. If you want more fog you can blow on the window." "I can write Karen's name." "Do I see some ice on the inside of the window? It must be really cold outside."

Save for Older Children:

Talking about humidity and moisture.

Using an indoor/outdoor thermometer and comparing the temperatures.

Eggs

You Need:

A raw egg, a hard boiled egg, colored candy eggs (optional)

Pictures of big, little and medium sized eggs and the birds that lay them*

Radish peels, onion skins, red cabbage or orange peelings for dying eggs

*Resource Book:

Burton, Robert, *Eggs: a Photographic Story of Hatching*

To Do:

When you are cooking or eating give toddlers an opportunity to see and talk about eggs. Show toddlers how eggs change. Eggs can be scrambled, hard boiled, deviled, poached, curried, shirred,made into custard, omelet, egg salad sandwiches, or eggnog. Whipping whites to a froth is fascinating.

For Dying: Boil peelings, skins or cabbage, each in its own pan, to get natural dyes. Reduce liquid. Initially, clean eggs in vinegar. For "resist" designs mark with light color crayon before dying.

New Words:

egg	hard/soft boiled
dye/color	fragile/breaks
scramble	raw
white/yoke	birds lay eggs

Save for Older Children:

Talking about and identifying eggs from different birds.

Toddlers still believe in magic, so Dr. Seuss' *Green Eggs and Ham* won't sound silly to them, but four-year-olds love the book.

Note:

Eggs are one of the most nutritionally complete foods and are inexpensive as well. We can usually find at least one way that toddlers will eat eggs. Even small infants can eat the yolk of a hard boiled egg. One of our favorites is "egg in a nest": Make a small hole in a piece of bread, toast one side of the bread in a frying pan or griddle, turn over and drop an egg in the hole. Put a lid on for a minute, remove, cook until the egg is toddler's favorite consistency (some like it soft), and serve.

See also Ornithology

Let's Go Outside

You Need:

An excuse to go outside

The proper clothes

Resource Books:

Angela Medearis' *We Play on a Rainy Day*

Alvin Tresselt's *Follow the Wind,* NY, Lothrop, and *Raindrop Splash,* NY, Lothrop

Let's Read and Find-out Science series *Feel the Wind*

Tomie de Paola's *The Cloud Book,* NY, Holiday House

Julia Hoban's *Amy Loves the Rain*

David McKee's *Elmer's Weather.* Colorful board book. Elmer is an elephant who encounters rain, fog, snow, lightning and wind.

To Do:

Get ready to go outside. Put on bathing suits for splashing in water and talk about how warm the weather is.

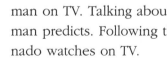

Put on snow suits and go play in the snow. Talk about why we need snowsuits and gloves.

Put on rain gear, find the umbrella and talk about why we need them.

Then go out, feel the warm sun, notice how hot the sidewalk is; feel the snow, taste the snow; look at the black clouds and taste the rain. Take an instant camera and bring back pictures.

TALK ABOUT ALL OF IT.

New Words:

warm/sun	hot/cold	wet/rain
clouds	splash	raindrop

Sample Questions and Comments:

"Let's go outside." It's nice and warm outside so we will want shorts." "Look at all that rain. We will need a raincoat and the umbrella to keep from getting all wet." "Doesn't rain feel good on your face? The rain tastes good too."

Save for Older Children:

Identifying the different clouds (see de Paola's book). Watching the weatherman on TV. Talking about what the weatherman predicts. Following the hurricane or tornado watches on TV.

More Air

You Will Need:

A plastic grocery bag for each toddler*

—*or*—

Crepe paper streamers

Rubber bands.

A windy day would be nice.

To Do:

Pull the bag through the air and catch some air. Show toddlers how to run with the "air trap." Tie a small length (never more than 18") of string to the air trap and run so that it becomes a kite.

Blow up the bags and close with a rubber band. Test the air in the bag with hands, Experiment with the "pillow of air." Toddlers will think of interesting things to do with them, like trying to pop them by stomping on them.

Cut streamers to about 20 inches. Twist one end as a handle and let toddlers run with them.

TALK ABOUT ALL OF IT.

Resource Books:

Gail Gibbon's
Flying, Catch the Wind, or *All About Kites*
Aliki's *Feel the Wind*

Sample Questions and Comments:

"Look how the air blows up the bag." "We've made an air trap. We'll tie this string to the handles of the bag and run with it. Watch how the bag traps the air and flies like a kite." "Now that we have trapped the air inside the bag, what do you suppose will happen if we sit on the bag?"

"What color streamer would you like to have?" "I will cut one for you." "I like the way the streamers wave in the air."

New Words:

air	trapped	poof
pop	blow	rubber band
bubble	escaped	air trap
sit	string	

Save for Older Children:

Regular kites, pinwheels, wind experiments, or talking about air currents and weather.

See also: Physics/Balloons and Natural Science/Air.

*Note of Caution!

Supervise carefully and toss bags in the trash when the project is over. Plastic bags can suffocate a small, helpless, and unsupervised child.

Rain, Rain, Don't Go Away

You Need:

A rainy day. Rain gear for adults and toddlers. A safe place to go for a walk. Puddles in which to splash or at least wade.

An instant camera—optional

See also Nature/Meteorology

To Do:

Go for a walk in the rain. Put your face in the rain for a minute and taste the rain. Watch water running in instant rivulets. Enjoy the adventure. Come back to a warm place with dry clothes, a warm drink and books about rain. Take a picture of the rainy-day adventurers and share it when you get back to home base.

Good Reference Books:

Medearis, Angela, *We Play on a Rainy Day*

Tresselt, Alvin, *Raindrop Splash*

Hoban, Julia, *Amy Loves the Rain*

Johnson, Angela, *Rain Feet*

New Words:

rain	wet	tongue
raincoat	umbrella, boots	
wind	water	gutter

Sample Questions and Comments:

"Lets get our raincoats on and go for a walk in the rain." "I wonder what rain tastes like?" "Oh, my face got wet. I'm glad we brought this little towel." "Look at all the water in the gutter."

Save for Older Children:

Meteorological talk—clouds types, predicting weather, hurricanes.

Tomie de Paola's *The Cloud Book* has very good descriptions of cloud types.

Shells

You Need:

A place where shells can be found—seashore, river, woodlands, or back yard if you are lucky

A pail or a bag to carry shells

Resource Books:

Eyewitness Books *Shells*

Irene & W.W. Robinson's *At the Seashore*

To Do:

Go exploring. Look for shells, collect them and enjoy them—talk about them. Put them in a special place to be enjoyed later.

Ask children to bring shells from home to share.

New Words:

shell   hunt   snail
snail egg/fragile

Questions or Comments:

"Lets go hunting for shells." "I wonder what kind of shell this is?" "Look at the crab carrying his shell on his back."

Save for Older Children:

Talking about and formally identifying different mollusks and reptile shells. Categorizing the shells will be fun later.

Note:

Crabs or snails are ideal animals to introduce "live" shell inhabitants. Garden snails can be the main attraction in a simple terrarium. You can even collect some snail eggs (they lay up to 50 pearly white eggs in a small hole in the soil—maybe you can find them under rotting leaves) so that you can watch them hatch. Keep the terrarium soil moist. Baby snails can eat lettuce. Put black paper over one side of the bowl to keep the light out. When you look under this you will see interesting things.

Historical Note:

Shells have been important to man as tools, decorations, jewelry, and money throughout history. A shell used as money by North American Indians was called wampum. Conch shells are still used as horns in some parts of the world. A large turtle shell makes a great percussion instrument.

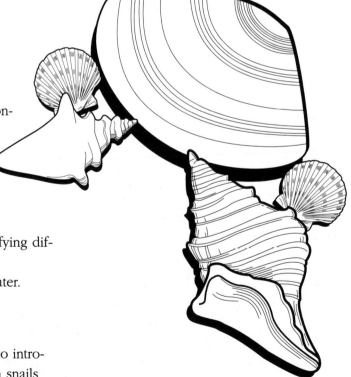

Explorers

You Will Need:

A sense of adventure

Time to explore inside—start by "child-proofing" a room or two.

Time to explore outside—yard, field, sidewalk, beach, park, tide pools, etc.

Resource Books:

Rhoda Readleaf's *Open the Door Let's Explore More!* Field Trips of Discovery for Young Children

Christopher Santoro's *Lift a Rock Find a Bug* A Chunky Flap Book

Chubby English/Spanish Books *Let's Take a Walk; Vamos a Caminar*

Sue Williams' *I Went Walking*

What To Do:

Say "Lets go for a walk."

Let the toddler lead you. Watch for serendipitous opportunities.

Use all your senses and show him what his senses can tell him.

Talk about what you see, hear, smell, touch, and sometimes taste.

Turn over a rock—what do you see?

Take a "new word" walk. Write down the new words.

New Words:

The list is unbounded:

Inside   outside
yard/field/neighborhood
walk/stroll/wander
exploring   names of things

Sample Questions and Comments:

"I wonder what is over here?" "What do you see?" "What is that I hear?" "What do you suppose that is?" "I have just the book at home. Let's look it up when we get back."

See also: Anatomy/senses/listening

Save for Older Children:

Identifying species, identifying sophisticated smells and using maps.

Nature

"The world's greatest language development kit is the world itself."
*—Ruth Wilson, "Nature and the Young Child: A Natural Connection"**

You Need:

A backyard, a school yard, a park, a national park, a patch of grass, a sidewalk with weeds growing in the cracks....

To Do:

Stir enthusiasm, focus attention, share inspiration

Look at the world through new eyes. Worry not about developmental domains and ponder more about enjoyment of the world around. Help the toddler see what he is looking at. Give him the gift of words. Watch the ants cart away food to their nest. Wonder about the wind. Let the toddler lead you. Enjoy.

New Words:

The list is endless:

listen	dig	poke
shake	touch/don't touch	pat
smell	try	

Sample Questions and Comments:

"Don't worry about naming everything—focus on beauty, mystery and wonder. Keep experiences simple."—Ruth Wilson

"Let's go up the hill and see what we can see." "What do you suppose would happen if...?" "I wonder what that is?" "Do you see the ladybug flying near the bush?" "The clouds† are very white today and they are moving fast." "I think it is going to rain." "Where did you find that bug?" "What kind of a bug do you suppose it is? Let's look it up when we get home." "How many sowbugs to

you see?"

Seeing shapes in the clouds is fun when you are four.

Tip:

If you plan to be outside with toddlers for a long period, especially during warm weather, be sure that you use some sunscreen—water proof, UVA/UVB protection, 15 or 30 SPF.

Note:

In nature you will find a lifetime of science and math curricula. Mother nature sets the stage for your serendipity. The new nature books that are available are awesome.

Joseph Cornell's *Sharing Nature with Children* is especially helpful for adults who want to share nature with toddlers. There are are over forty glorious *Eyewitness Books* alone. Dorling Kindersley's *My First Look at Nature* with its vibrant illustrations divided into seven groups—flowers, minibeasts, trees, animals, water life and seashore life, is very helpful. The *Usborne Complete First Book of Nature* and Wee Pudgy Board Books *Wee Wonders of Nature* are also beautiful publications as is Sue Williams' *I Went Walking*.

Young Children, Sept. 95, pp. 4-11.

The Unexpected

You Need:

An unexpected event that interrupts your plans.

Resource Book:

Joseph Cornell's *Sharing Nature with Children*

To Do:

Look around you and see what is happening. If workmen have had to come into your classroom, watch them and talk about it. Follow-up by drawing pictures of them, pretending to be them or making up stories about them. Maybe you can find some music that is appropriate. Go with the flow—catch the serendipity. Do what you planned for today another time.

If the weather spoils your plans, use the weather. Watch it, talk about it, and experiment with it. Read books about it and sing about it. Mother nature is giving you the best curriculum materials there are.

New Words:

surprise/taken by surprise
disappointment
change of plans   helper

Sample Questions or Comments:

"What's going on here? Look at the water coming in. The roof is leaking. We will call someone to fix the leak." "The rain water left a big wet circle on the table. Let's get some towels and dry the table. Then we can put a pot under the drip so the table doesn't get any wetter." "I wonder what will happen when…?" "What will we do now?"

See also: NATURAL SCIENCE/Meteorology/Rain, MEDICINE/emergencies

Getting Ready to Get Ready to Read

You Will Need:

Colorful but expendable mail order catalogs—toy catalogs are good, but any catalog that has pictures of interest to toddlers, i.e. babies, dogs, cats....

To Do:

Play and learn Hap Palmer's "Come Read a Book" from his cassette/CD *Babysong*.

Use catalogs to expose young toddlers to short cozy moments of reading. These expendable materials are great lap books, as the toddler points out what she sees. Read product descriptions so that she can hear the language.

The toddler can have a book of her very own. Worry not about the catalogs being mangled. Then after young toddlers have had lots of experience grasping and turning clumps of pages, real books (board books at first) can be handled with more care.

New Words:

book   catalog
names for objects   my own

Note:

Help toddlers focus on distinct items on the page. Show them how and which way the pages turn, and point out that new and exciting things appear on following pages.

Save for Older Children:

Choosing things from the catalog that they want to buy.

Naming Objects

You Will Need:

No materials needed

Resource Books:

Toddler Time™, Thinking, Little Explorers, I can Do It!, Numbers, Sorting

To Do:

Play "Lets Look out the window and see what we can see."

Read and ask what the toddlers see.

Take a walk (in the room, outside, at the market…) and see what we can see. Talk about it.

Sample Questions and Comments:

"Let's see what we can see." "I see a car?" "Do you see the car?" "What else do you see?" "Do you see the grass?" "The grass is the same color as Kevin's shirt. It is green isn't it?" "What is that?" "Yes, what does the cow say?" "I wonder what that is? Mary says it is a flower. I think she is right."

New Words:

Names for objects:

car, grass, tree, bird, cow, toaster, chair, etc.

Note:

Teach a toddler one item at a time. Describe the object if she doesn't see it at first. Go outside and touch the objects that you have just named, if you can. It is interesting to have a toy, a picture and the real thing at hand at the same time—also recognizing something on TV and then in real life, can be a new experience for a toddler.

Save For Later:

Activities that identify unusual items—e.g. telephone wires, underground utility box. Talking about makes of automobiles, kinds of bikes, or other complicated abstractions. The "Look Out the Window" game is fun to try each of the four seasons—things look very different.

Birdfeeder

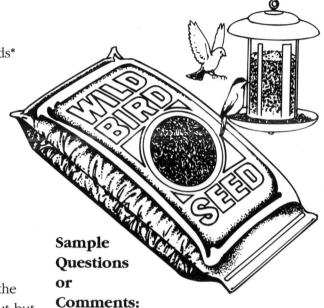

You Need:

Bird seed

Large pictures of common feeder birds*

Peanut butter and plastic knives

Dry cobs of corn or pine cones

yarn or string

Resource Book:

Martchenko, Michael Martchenko's
Birdfeeder Banquet

To Do:

Give a toddler a chance to examine the pine cone or cob. Let him spread peanut butter on the cob or pine cone. Then roll it into the bird seed. The toddler will enjoy nibbling on the peanut butter. The birdseed is fascinating to handle, pour and run through one's fingers. Although not recommended, birdseed is ok to eat. Do this project outdoors so that the birds can eat whatever is spilled. Hang the "birdfeeder" where the birds can be safe from cats.

Birdfeeders are much easier to maintain than caged birds—no cage to clean, no vet bills. You can increase the number of visitors by adding different seeds, e.g. thistle seed attracts gold finches, peanuts attract jays. The feeding station can be as simple as a hanging a plastic bottle with the sides cut open.

New Words

bird seed	peanut butter	pour
spread	birdfeeder	corncob
roll	coat or cover with seed	

Sample Questions or Comments:

'What do you suppose birds like to eat?" "Lets make a birdfeeder." "I will help you." "This is special seed just for birds." "They like peanut butter too."

Save For Older Children:

Talking about and identifying some of the world's 8000 species of birds.

Shelling peanuts for the birds and making 1/2 of them into peanut butter in a blender to put on the feeder above.

Note:

A toddler may not understand what is happening yet, but will enjoy the process, nibbling the peanut butter as he goes. As he watches the birds eating from the feeder he made, he will get the idea.

*One source of excellent bird pictures is the Cornell Lab of Ornithology—Project Feeder Watch, Ithaca, NY.

Birds

You Need:

A place where birds can be observed*

Large pictures of common birds†

To Do:

Sit as quietly as your toddlers are able and watch birds in their natural habitat. Some birds (ducks for instance) love to be fed. Talk (whisper) about it. Birds fly, swoop, flit, jostle each other. Talk about it.

Read *Are you My Mother?* and look at the pictures of birds, and share Gillimard Jeunesse and Pascale de Bourgoing's First Discovery Book – *Birds* and Alexandra Parsons' *Amazing Birds*.

New Words:

bird   seed/insect
sing/tweet/cheep
flit/fly/swoop
jostle   chicks

Sample Questions or Comments:

"I see some birds (whisper). Lets watch them for a while." "What do you suppose the birds are eating?" "I like that bird's song." "Is that a fish I see in the eagle's claws?"

Save for Older Children:

Talking about and identifying some of the world's 8000 species of birds, from the tiny bee humming bird to the 11.5' wingspan albatross.

Spring nest building, eggs and chicks will be of great interest later.

Have you ever seen a raven or gull drop a shell to break it open?

*Note:

Birds can be found everywhere—gulls at the beach, ducks on the pond, eagles at the mountains, ostriches at the zoo, sparrows at the fast food patio, finches at the feeder. Specimens that don't fly away can be found at the Museum of Natural History, along with samples of their nests & eggs.

†Large colorful pictures of common feeder birds are available from Cornell Lab of Ornithology.

Scavengers: Nature's Custodians

You Need:

A place where birds can be observed scavenging road kill. Ravens and vultures are often seen cleaning our roads.

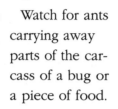

Watch for ants carrying away parts of the carcass of a bug or a piece of food.

Resource Book:

Parsons, Alexandra, *Amazing Birds*

To Do:

Find an observable carcass, point out what is happening, stand nearby or park a vehicle where toddlers can quietly watch for scavengers. Follow the ants or beetles to their home, and talk about how they help clean up as they take the food home for their families.

New Words:

carcass	ants	vulture
raven or crow	dead/carcass	

Sample Question or Comments:

"What do you suppose birds like to eat?" "That bird is taking food back to their nest. He is cleaning up the road for us."

Save for Older Children:

Be ready to get roped into discussing dead animals and why they die and why scavengers eat the dead animals.

Talking about and identifying species of scavengers—vultures, crows, ants and catfish, who are the aquarium's "vacuum cleaners."

The burying beetle digs soil from under a dead animal, so it is slowly buried, then lays eggs on the body, so its young can feed off the body when they hatch.

Note:

The toddler may not understand what is happening yet, but may enjoy the process. Think about or take along something for him to do while waiting or watching.

Blocks/Ramps

You Need:

A handicapped access ramp. There are many around these days.

Triangle blocks, boards and blocks, balls, little cars

—or—

Vinyl-covered play foam ramp from a toddler play center

To Do:

Turn toddlers loose on a safe handicapped access ramp. You don't have to tell them what to do. They will climb up, they will run down. They will invent things to do with the inclined plane that you never thought of. Take a ball to introduce later to extend the play. Call the ramp by its names—"inclined plane" as well as "ramp." Then, as you look around you, you will notice that our slide is an inclined plane and that we have blocks that are ramps.

Show toddlers how to build a ramp with blocks. Show them that if the ramp is steep even dirt and sand will slide down. Show toddlers that a ball will roll down the ramp (inclined plane) and that little cars will also go down a ramp.

Use the covered foam triangle as the center piece of play. Show toddlers that balls will roll down the ramp. Get an adventurous toddler to somersault or roll down the ramp.

New Words:

ramp	slide	inclined plane
car	ball	down
roll	board	bump

Sample Questions or Comments:

"I wonder what will happen if we put a ball up here on the top of the (ramp) inclined plane?" "Let's build a ramp with blocks. It is also called inclined plane. Cars roll right down it."

Save for Older Children:

Talking about why we have access ramps, and how much easier it is to roll wheelchairs down ramps instead bumping down steps or curbings.

Competitive racing of cars down the inclined plane.

Galileo, mankind's first scientist, used the inclined plane as he began discovering the laws of nature.

Cooking

You Need:

Any frozen food that can be eaten frozen—peas, grapes, bananas, berries...

A duplicate food that is thawed or cooked.

Two tubs of water—one nice and warm, the other with ice cubes floating in it.*

Resource Books:

Fit-A-Shape™ *Opposites,* and Rainbow Books, *Opposites*

To Do:

Serve food in bright colored paper or plastic cups.

Talk about the cold and hot food.

Enjoy and talk about it.

Next time you need water to clean up, introduce two tubs, and talk about who wants to wash up in which tub.

New Words:

peas   grapes   banana
cold   frozen/cold/hot/warm
cooked

Sample Question or Comments:

"Let's taste frozen peas." "I wonder what they will be like." "They are cold, aren't they?" "What do you suppose cooked peas will taste like?" "I like the cooked peas better than frozen ones."

"You can use whichever tub you like to wash." "They are different, aren't they? Why do you like that tub of water better?"

Save for Older Children:

Waiting to freeze food. Making juicesicles.

Making homemade ice cream.

Chopping the ice, tasting the salt, taking turns cranking the ice cream maker—toddlers do like to watch, though.

See: Physics/cold/freezing

See also: Math/opposites and Nutrition/fiber

*CAUTION:

Never leave buckets of liquid unsupervised. Young toddlers (to 15 mos) have been known to drown in even small amounts of liquid.

Note:

Green peas have three times more fiber, 40 percent more iron, and 25 percent more vitamin A and vitamin C than green beans. And best of all, toddlers seem to like them best.

Evaporation

You Need:

A blackboard or

A warm sidewalk

Water

a brush* or a finger to dip into the water

Resource Book:

Rainbow Books, *Opposites,* colorful, inexpensive board book with lively photographs

To Do:

Find an opportunity to talk about evaporation. Point out opposites whenever you can.

New Words:

evaporate	paint	blackboard
sidewalk	wet & dry	warm
brush/big/little/sponge brush		

Sample Questions or Comments:

"Look! I can mark the blackboard with water. Let's write Karen's name." "Oh, it is evaporating." "Where did the mark go?"

"Water evaporates from the blackboard." "Where did it go? Into the air."

Fun to Save Until Later:

Older children will eventually notice that evaporating makes things colder—then is the time to talk about refrigeration and how it works.

See also: Physics/air & Physics/more air and "Anatomy/Feetprints"

*Note:

"Painting" with water can be one of the most popular activities for toddlers on a warm day. You can use big brushes, small brushes and/or sponge brushes. They can "paint" bikes and walls, as well as sidewalks. A "car wash" is always fun.

Freezing & Melting

If you are lucky enough to have one or two hand-crank ice cream freezers, you can keep toddlers and other children, for that matter, busy most of a day (and yourselves as well) making ice cream (frozen juice, frozen yogurt, slush…)*

You Need:

2 blocks of ice (one for playing and one for use)

Large bag rock salt

Ice cream freezer (or two for a large group)

Big bowl and spoons

    4 eggs, beaten

    Dash of regular salt

    2 c. sugar

    2 qts. milk

    1 1/2 c. whipping cream
       or 1/2 and 1/2 for lighter ice cream

    3 T. vanilla

    Small dash of lemon flavor

To Do:

Let toddlers help chop the ice. Pour some of the rock salt on the ice and watch it eat into the ice. Do not let a toddler eat the salt.

You can add food coloring to the block of ice later to extend this activity. Talk about melting and freezing. Put a small piece of ice on the cement and watch it turn into water.

Have cooks wash their hands. Let toddlers watch you measure the ingredients into the bowl.

Beat eggs. Add sugar and salt, and stir until dissolved. Add the rest of ingredients. Let toddlers stir. They can watch you pour the mixture into the freezer, close up the lid, and can help add the chopped ice and rock salt. Be prepared to do most of the cranking, but toddlers will want a turn. You can use an electric freezer if time is short.

New Words:

ice/ cold	frozen	freezer whip,
stir	rock salt	melting

Sample Questions and Comments:

NOTE: Toddlers will not grasp the chemistry of the brine or the physics of the equilibrium between the ice and the cream, but will enjoy the process many times.

*Simpler recipes for slushes, ices and other ice creams are in almost any cookbook.

Save for Older Children:

Talking about the freezing process. Sugar crystals and rock salt look similar. Taste the salt. Taste the sugar.

Making Noises With Paper

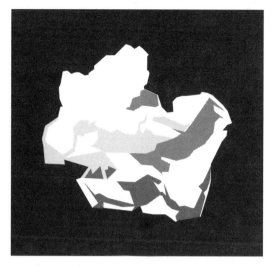

You Need:

Large cardboard (a back of a large sketch pad will do)

Medium size cardboard (back of a canary pad)

small cards for each toddler (3x5 or 5x7 cards)

Sheets of newspaper

Resource Book:

Usborne Books at Home, *Who's Making That Noise?*

To Do:

Take large cardboard between two hands. Then flex and bend back-and-forth fast until you get a "boing." Repeat with each size board. Give toddlers cards and let them try. Even fanning the cards in the breeze will result in a little noise. Each size makes a different sound. Talk about it.

Try making the noises to music. The large piece of cardboard sounds a little like an Australian musical instrument.

Give each toddler a piece of newspaper and help her crumble it. Paper makes lovely crunching and tearing sounds. Put the crunched paper into the recycling bin.

New Words:

cardboard card little/big
back and forth crush, crinkle, tear
boing

Sample Comments or Questions:

"Listen, what is that noise?" "I wonder what noise this piece of cardboard will make." "Here are some cards. Let's see what noise we can make with them." "This little card makes a little noise. We have to listen carefully to hear such a little sound."

Note:

A saw or thin plate of metal will make a very satisfying "boing." If you have someone who can play a saw, lucky you!

Mirrors and Mirroring

You Need:

A mirror large enough for all the toddlers to see into at the same time.

A small mirror that can be held by the adult

A flashlight

To Do:

Stand in front of the mirror and talk about what you see. Can the toddler identify himself? His neighbor?

Ask where parts of their bodies are and have them point to them.

The mirror is confusing at first—be sure this activity is fun.

Play in front of the mirror.

Use the hand mirror to reflect the sun. Move the image and talk to toddlers about what they are seeing. Let toddlers see themselves in the little mirror. Try using a flashlight and watch the reflection dance—let toddler follow it.

Note and talk about the symmetry and reflection patterns.

New Words:

mirror   reflection   light

Sample Questions and Comments

"I see two Roberts. How can that be?" "Can you touch your reflection?" "The light comes back. Its reflection is touching the floor. Can you touch the reflection of the light?"

Note:

We all learn from others as they "mirror" our expressions. Infants and toddlers are especially inspired as we mirror their facial expressions.

Save for Older Children:

Trying to button while looking into a mirror. Trying to write our name while looking into a mirror. Trying to draw a picture while looking into a mirror.

Noises

You Will Need:

Pictures of things that make noise

Items that make noise, like a cooking timer or alarm clock, bells

—*or*—

Time for a listening walk.

Musical Resource:

Tom Glazer's *More Music for 1's and 2's.* Cassettes for the very young including "What does Baby Hear?"

Good book:

Usborne Books at Home, *Who's Making That Noise?*

To Do:

Talk about the noises that things (or animals) make. Most toddlers know that a cow "moos" and a rooster goes "cock doodle do" etc. Try train noises and other unusual things.

Sample Comments or Questions:

"What is this?" "Yes, this is a cow. What kind of a noise does a cow make? Does she sing?"

"Look what I have. It is a clock. It tells us when it is time to eat or wake up, or go to school." "Lets listen to the clock." "Can you hear it with your ears?" "Where are your ears?" "What is the clock saying? To me it sounds like it is ticking—tick tock" "Listen, what do you hear? I think I hear…"

New Words:

noise/sound   names for things names for the noises they make descriptions of the noises

Note:

Teach toddlers one or two noises at a time. Describe the object and make its noise. Take a listening/noise walk. Cars honk their horns, birds sing, the wind whistles through the trees…you may even be lucky enough to hear a siren. Children giggle, sing, and shout. Animals bark, growl, mew, or cackle. Machines vroom, putt-putt, tick-tick or ding-a-ling.

Later:

Talking about the different noises that the same animal makes. People make a lot of different noises (burp, sneeze, laugh, whisper, etc).

"Ooh! Ooh!"—Toddler is Testing Gravity Again

You Need:

A safe place for toddler to test gravity—a highchair has often served this purpose.

A basket or other container to retrieve dropped items

A variety of items to drop—heavy, light, long, short, round...

Useful Resource Books:

Williams, Rockwell, & Sherwood's *Mudpies to Magnets: A Preschool Science Curriculum*, and *More Mudpies to Magnets: Science for Young Children*.

To Do:

Make a place for toddlers to drop things to the earth—one of their favorite activities (Drop not throw). Provide each toddler with a variety of things—heavy things, middle sized things, a feather, and talk about dropping things. A feather and a sheet of paper float to the ground. A ball drops faster.* What happens when you drop a rock? Getting off the platform and retrieving everything will be interesting too. Be sure to keep others from the target area. Later, you can also use the basket for toddler to "score points."

There is a law that says that everything is either at rest or in motion. A toddler is mostly in motion and is dropping things constantly, so these experiments, and the discussion thereof, can go on often. Don't forget to play "Ring Around the Rosy." Toddlers love to "all fall down."

New Words:

drop   gravity   feather
sheet of paper   float   heavy/light

Sample Comments or Questions:

"What will happen if I let go of this rock?" "Will the feather drop to the ground too?" "It looks to me like gravity is pulling everything to the ground." "Uh oh! Jerry just lost his battle with gravity. I wonder why he didn't fall up?"

Save for Older Children:

Talking about Galileo's question—"Why do things drop to the earth?" Why don't they fly to the sky? Find and read "Falling Up."

*Note:

Young children will not know or care about vacuums and their effect on gravity nor Newton's and Galileo's laws.

Push and Pull

You Will Need:

Two skateboards, skooter boards or bases with wheels that move freely

What to Do:

Have a toddler put the wheel toy next to a wall with her feet up on the board. Show her how to push off from the wall.

Have toddlers sit on the wheel toys opposite each other, touch each other with their hands or feet, and talk about what is happening. As they interact they will move in unexpected ways. The adult may have to demonstrate.

New Words:

skateboard   wall   push/pull
wheel toy   Action/reaction

Sample Questions or Comments:

"I wonder what would happen if…" "Let's try to…"

Tip:

It is hard to arrange, but sitting on a wheel base and trying to push a swing brings about an interesting experience—could be dangerous, though.

Save for Older Children:

Discussing Newton's Third Law: for every action there is an equal reaction.

Sticking Things

You Will Need:

Sticky things: cellophane tape, magic tape, masking tape, mailing tape, plumbers tape.

To Do:

Cut a few pieces of each kind of tape and tip them on the side of a child-size table. Stick a piece in the palm or on the back of toddler's hand. See what happens. Add sheets of paper and show each toddler how to make a collage with the tapes. Talk about the different kinds and colors of tape.

New Words:

tape   stick   sticky
name of tape

Sample Questions and Comments:

"Look at the different sticky tapes we have." "This is called Scotch tape." "Let me give you a piece of tape. It feels sticky, doesn't it?" "This one is plumber's tape. It is used to keep pipes from leaking." "I wonder what would happen if we…" "Let's make a collage. We can take it home to show your Mommy."

Save for Older Children:

Making mosaic collages with small, cut pieces of tape.

Gluing projects: the toddler is more interested in pouring glue than in making artistic collages.

Using sticker books and collecting stickers: if you introduce stickers too early, the idea of the stickers—except for stickiness—is lost on toddler. Then when stickers could be used as an educational tool they are considered "baby stuff" by older kids.

Water

Most Early Childhood classrooms have ongoing water play. If you do, take a look at the setting. Eliminate the clutter and introduce one tool at a time to a toddler. What is new here is that we suggest that water play be taken more seriously, exposing older children to simple experiments, discussions and maybe even recording some experiences. Discuss what is taking place and wonder what would happen if…. The toddler will go along with whatever happens—she loves water play and will imitate what she sees.

You Will Need:

A source of water   A hunk of ice

The usual water play tools and water table when appropriate.

(Thrift shops and "surplus" educational materials companies have wonderful water play supplies)

At least 1 one-gallon plastic bottle—one per toddler is better

A hot sidewalk or warm surface that will dry fast—paintbrushes

A hose with pressure nozzle (for the adult)

To Do:

Enjoy: Richard Sharp & Norma Seymour-Simon's *Wet World* & Mike Thaler's *In the Middle of a Puddle*.

Water play can help introduce the three states of matter: liquid, solid and gas, and the concept of change. In its liquid form water can keep toddlers busy for hours. Just ask any nursery school teacher. Our job is to provide a vocabulary for all the wonderful things that happen to water.

Ice is cold and it melts.*

Water painted on the sidewalk evaporates (see also evaporation).

Water poured into a funnel comes out the bottom. When sucked through a straw, water climbs up into your mouth. A big (1 gallon) light plastic bottle is very heavy when filled with water. (Let toddler hold the bottle while you fill it—it gets heavier and heavier). It is fun to lug.

Fast moving water can move light things like leaves, but may not move heavy things like a brick.

New Words:

The list is endless: water pour rain/shower ice slush evaporate/dry drip/drop melt suck paint heavy overflow water pressure splash leaves

Sample Questions and Comments:

"How many ways can you think of to carry water?" "Can we carry it in our hands?" "What do you suppose will happen…?" "The water that we painted on the sidewalk has disappeared. It evaporated." "The water evaporates. Where did it go? Into the air." "This ice is cold. Let's put a piece of ice on the warm sidewalk and see what happens."

*Save for Older Children:

Water takes the shape of whatever container it is in. Some things float on water, some do not, a notion that is lost on toddlers. Waiting to see something freeze is hard for toddlers. Water turns to steam in your humidifier. Watch for it when you take a hot bath. Watch for the mist-steam-fog coming off the meadow on a warm winter morning. The water evaporates into air to become clouds.

Note: Water is probably a toddler's favorite plaything. Water will soothe an upset toddler. He will stand on a stool eternally to play with even the smallest dribble of water.

By the way, water is our best beverage. Toddlers love to serve themselves from a jug with a sigot.

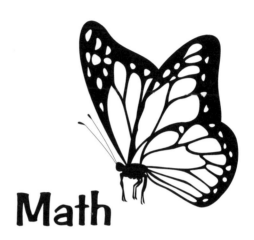

Math

"To those who do not know mathematics it is difficult to get across a real feeling as to the beauty, the deepest beauty, of nature."

Richard Feynman, The Character of Physical Law

Reader's Objectives

_____
_____
_____
_____
_____
_____
_____
_____
_____
_____
_____
_____
_____
_____
_____
_____
_____
_____
_____
_____
_____
_____
_____
_____
_____
_____
_____
_____

Introduction to the Math Activity Pages

DEVELOPING MATH CONCEPTS

"Pure mathematics is, in its way, the poetry of logical ideas."—Albert Einstein

As you participate in the following math activities keep in mind that the toddler's early math learning must be simple, basic and personally relevant. Because concept formation cannot be forced or "taught," adults can only help toddlers pass through their own pre-programmed stages of concept development. These stages are related to the child's chronological and mental age and experiences.* So don't limit yourselves just to the activities listed. Use the Activity Pages to open your mind to experiences especially relevant to your child.

Unlike language development, we do not notice mathematical growth because we think mathematics is only learned in school. Not so—writing numbers comes last. First come the ideas built from toddler's experiences with his world and the objects in it.

Our job then is to lay the foundations for concept development with a multitude of daily practical experiences combined with informal talk, e.g. "Would you like two cookies, one for each hand?" So, set the stage and let toddlers

experience the serendipity—the "happy accidents."

It is interesting that scientists have discovered human learning is not necessarily enhanced by drilling of isolated academic skills. What we ultimately want is children who possess thinking, reasoning, decision-making and problem-solving abilities. There is the challenge. Toddlers will still insist on repetition—practice, practice, practice—but on their own terms and at their own pace.

Measuring, ratios and scale

Because toddlers see the world through their own eyes, they, with so few experiences under their belt, learn concepts in relation to themselves. A toddler's early measuring often uses her own body as reference— "Let's build it as tall as I am" or "It is as big as my finger" or "Daddy is bigger than I am." A wide open arm span tells us she loves us "this much"—a perfectly acceptable unit of measure.

We want to establish the notions of ratio and scale early in math learning. Our society suffers from innumeracy and, as a result, we

*See the Appendix for an informal description of what to expect during the Ages & Stages of toddlerhoods.

tend to be penny-wise and pound-foolish. Understanding ratios—"How big am I compared to Daddy?" and understanding scale—"How many of me would it take to make one Daddy?" are important. Both notions can be illustrated visually. Putting numbers to these concepts can come later.

Make an effort to talk about comparisons—tall/short, few/many, less/more, full/empty. There are a number of good children's' books devoted to these comparisons.

Sorting and Categorizing

As you put toys away you can sort and categorize: "The blocks go in the block corner." "The long blocks belong here." "Let's put the ramp blocks together here." "The teddybear counters go into the basket." "I see two more puzzle pieces on the floor. They belong in the basket. The basket goes on the puzzle shelf."

Music, crafts and Math— Symmetry and Reoccurring Patterns

Scientists at University of California Irvine have done some interesting musical experiments that "dramatically improved children's spatial reasoning." They say that music stimulates inherent brain patterns and enhances complex reasoning tasks. We do know that toddlers love music and will dance and sing from the minute they are on their their feet. Howard Gardner says that music is the first mode of learning. Remember how you learned the ABC's? Throughout the Activity Pages you will find music ideas and the Bibliography includes some favorite toddler music.

Expose toddlers to rhythms and patterns

through music. The pattern of chanting and clapping hands is a good way to start. One does not need a singing voice to chant and we all can clap.

As children get older they can experience symmetry and patterns while making paper chains, beading, weaving and observing the beauty of nature's symmetrical patterns.

Spatial relationships

Toddlers love to dump or pour, drop things, crawl under, climb on top, or hide behind. They are absorbed with the notions of inside, outside, up, down, over, under, above, below and behind, even though they don't know the words yet. You can help toddlers label and enjoy these pre-math ideas as they experience them.

Object permanence

Peek-a-boo and hide-n-seek are good games to play with infants and toddlers. Each variation of these games establishes some idea of object permanence as well as addition or subtraction.

The Block Builder Mathematician

Our little experimenters can get lots of pre-math practice as block builders and we can learn about our toddlers by observing or joining them in their block play. They start simply—one block upon another. With your encouragement the toddlers will build more and more elaborate constructions. Try keeping a picture record of their efforts and watch how their skills grow. Blocks also provide opportunities to experience width, height, depth, nearness, separation and order.

Similarly, inherent in the use of blocks is the knowledge that some things are smaller, larger, longer or heavier, half the size of another, or twice the width of another. Using blocks also gives the child a chance to know and talk about squares & cubes, rectangles, and triangles.

Shapes

An intriguing stage of toddlers' development is the ability to "draw" circles. After a toddler learns to hold a crayon or pencil, he scribbles—mostly up and down—until one day the idea of circles and wavy lines comes into play. An adult can facilitate this learning by helping toddler "draw" big circles in the air and large circles on paper.

I remember my grand-toddler-in-arms reaching for the full moon one lovely night. "Mine, he said, "My moon!" We DON'T want to discuss how civilization learned about π, or the diameter of circles, to broaden our toddler's mathematical horizons, but we can introduce the words "sphere" and "circle" into a toddler's vocabulary, play with balls of all sizes, and enjoy finding circles wherever we can in a toddler's world.

Counting

Learning numbers is much more interesting than merely counting. The ideas of more, less and the same are much more basic. One of the first words a toddler learns is "more."

Although rote counting with rhymes and songs is fun for the toddler, the concept of numbers should be approached gradually through experiences with numbers of fingers, toes, buttons, cookies, pegs, or beads. Be aware though that young toddlers will be more interested in pouring the beads or teddybears from their container than using them as counters. Later, children will be using a number of colorful items as counters, including bears, stones, and shells. These manipulatives, including math rods, have been used by Montessori-type schools to help children visualize math concepts. Here again, blocks are a perfect manipulative medium for early experiences with numbers.

Save for Older Children

Toddlers are not ready for learning units of measure, concepts of fractions, complex comparisons, orders of magnitude, or the concept of area. These are ideas that should wait until the child is older. Complicated classifying, the idea of odd and even or word puzzles are things you can introduce later. Graphing, logic, multiplication, division, weights and the art of estimating, are all fascinating areas of math waiting for the older child. Learning to tell time must also wait for a year or two.

Legos™ and other building toys will give older children delightful opportunities for further math and spatial learning. Older toddlers may start enjoying the toddler version of Lego called Duplo™

Circle Fun

Materials:

Round things nearby

Foods that are circles; e.g. crackers, slices of orange or banana (try Vicki Lansky's *Slippery Circles,* finger Jello in a small OJ can with banana in the middle—see pg. 43 of Vicky Lansky's *The Taming of the C.A.N.D.Y. Monster.*)

Crayons and large paper

Resource Books:

Catherine Ross' *Circles: Fun Ideas for Getting A-Round in Math*

Judy Feldman's *Shapes in Nature,* Eyewitness Juniors

Activity:

Talk about round things. Look for circles around you.

Eat and talk about round crackers, Cheerios, M&Ms, etc.

Make big circles in the air and then draw circles on paper (nice big ones)*

If in a group, form a circle and play a game like the Hokey Pokey—talk about how we make the circle of people. Talk about sitting in a circle at story time.

Share the pictures and the silly words in Dayle Ann Dodds' *Wheel Away!*

New Words:

Round	Circle	Cracker
Shape	Draw	

Sample Questions or Comments:

"Can you draw a big circle in the air?" "Can you make a circle with your mouth?" ("oooh") "Can you make a circle with your arms?" "Can you make a circle with your fingers (demonstrate)?" "Let's hold hands and make a circle of people." Enjoy talking about numbers: "You have three round crackers, one for each hand, and one in your mouth."

See also: Geometry/More Circles, Nutrition/Pancake Circles, Geometry/Eating Shapes

Save for Older Children:

Talk about cylinders & spirals (regular and Archimedean). Have you noticed that you can cut a spiral as you peel an orange or apple? Egyptians used ropes and knots to measure circles; talking about circles and the magic number 6; counting the seconds by saying "one gorilla, two gorilla…" Talk about pebbles making ripples when dropped into water.

*Note:

Toddlers will be developmentally ready to graduate from drawing up and down scribbles to making circles. Be sure to allow uninhibited drawing of circles. Squeezed and tortured "pictures" will take the joy out of the experience.

*See the Appendix for an informal description of what to expect during the Ages & Stages of toddlerhood.

Eating Shapes

You Need:

Crackers in a variety of shapes: triangles, squares, rectangles or circles. Cut cheese or lunch meat in the same shapes (with cookie cutters or by hand).

Useful Resource Books:

Judy Feldman's *Shapes in Nature,* Eyewitness Juniors—a little old for toddlers, but one of a few books that begin to address simple math concepts

Fit-A-Shape, Quarto Children's Press

Snapshot™ Books, *Shapes Galore*

To Do:

Talk about the shapes as you let toddlers watch you cut and prepare. Then talk about matching the shapes as you serve and eat them.

Older toddlers may match the shapes as they serve themselves.

You can extend talk of shapes by using circles (English muffins, bagels, Cheerios®), triangles (tortilla chips, apple or pineapple wedges), or squares or cubes (brownies, waffles, lasagna).

New Words:

cheese	triangle	cracker
square	circle	rectangle
cutter	match/same	

Sample Questions or Comments:

"Which shape of cracker would you like?" "Would you like some cheese/meat to match your cracker?" "I think I will eat a rectangle." "The squares are my favorite."

Save:

Testing the children to see who knows what shapes.

Note:

The next time you have sandwiches you can ask your children how they want their sandwiches cut—into a triangle or a rectangle. You can even make a sandwich into four squares or four triangles.

More Circles

Materials:

A stick with which to draw a circle

Enough pebbles to make a small circle in the dirt

Hula hoop or a bracelet

Paper and crayons for paper, or chalk for the sidewalk

Wheels & tires are circles

Round food (crackers, cheese, cookies)

Resource Books:

Ross, Catherine Sheldrick, *Circles: Fun Ideas for Getting A-Round in Math*

Judy Hindley's *A Piece of String is a Wonderful Thing*

Activity:

Draw huge circles in the air.

Draw huge circles on large paper or on the cement with chalk, in the sand or dirt.

Make a small circle of pebbles.

Plan circle and sphere day—look for circles and spheres all around you.

Share Dayle Ann Dodds' Wheel Away! Enjoy the silly word play and find the circles.

TALK ABOUT ALL OF IT.

Sample Questions and Comments:

"Let's make circles...in the sky...in the dirt...on the paper...on the cement." "Can you make a circle with the rope?" "How would you like to eat a circle?" "Let's shuffle our feet in the sand and go around in a circle and see what happens." "Does anyone see any other circles?"

New Words:

Rope   Circle   Round
Tire/Wheel   Swing   Cracker

Save for Older Children:

Talk about the difference between circles and ovals. Talk about how the ancient Egyptians used rope and knots to measure and make squares. Play the scribble game—the child makes a simple scribble and the adult makes a "picture" out of the scribble and they talk about it.

See also: Math/Geometry/Circle Fun, Nutrition/Pancake Circles

Note:

Toddler will be developmentally ready to graduate from making up and down scribbles to making circles.

Spheres

discuss. Show them how each bubble frame works. The youngest toddlers will be happiest popping the bubbles. Talk about the recipe and measurements.

Make bubbles and, interchangeably, call them spheres. Enjoy.

Look for other spheres around you. Enjoy talking about circles and spheres when the opportunity arises.

Sample Questions or Comments:

"Lets see if this string will make a bubble." "How about those bubbles!" "Lets try a berry basket." "Let's swing the basket in a big circle." "Oh, the basket makes a lot of little bubbles." "Jason likes to pop the bubbles." "Blow on the bubble and see what happens." "What a big sphere that bubble is!" "The grapefruit we had for snack was not as big as that bubble." "Lets sing, 'I'm Forever blowing Bubbles.'"

Neat Things to Save for Older Children:

Learning to dip and blow their own bubbles; experimenting with different bubble blowing instruments—food grater, funnel, fingers!, hula hoop…

Talking about surface tension and how it helps confine the air of the bubble.

Note:

Young toddlers will not be able to blow their own bubbles, but will love to chase and pop them. Older toddlers may be able to blow lots of bubbles in the bottom of a cup without sucking…a nice noisy activity. Beware of soap in people's mouths, though.

You Will Need:

Bubble making paraphernalia

Bubble instruments: plastic cup with the bottom end cut off—inside another cup; a string tied into a circle strung with 2 1/2 straws for holders; plastic circles that hold a six-pack together, a berry basket (especially easy for toddler to swing), or even a plastic circle from a commercial bubble set. Bev Bos uses a hulahoop dipped into a child's wading pool.

Spheres—ball, orange, balloon, grapefruit (Bubbles are, after all, spheres that can be made with circular bubble frames)

Angela Webb's Talk About Book *Air* has some good ideas for bubble activities.

New Words:

round	circle/sphere	orange
bubble	pop	chase
balloon	float	

What To Do:

Mix the bubble solution while toddlers watch. Have an array of bubble blowing equipment available for toddlers to touch and

More Spheres

You Will Need:

A variety of balls: basketball, tennis ball, golf ball, marble

A milk carton or other cubic shaped item

To Do:

Take the opportunity to talk about spheres while having fun with any ball. You can also talk about big and small things.

Try kicking a ball to see how far it goes. Try kicking a milk carton. How far does it go? Does it roll?

New Words:

sphere/ball	big/small	roll
kick	milk carton/rectangle cube	
far		

Sample Questions and Comments

"Today we're going to play with spheres." "Yes, you're right. This is a ball. It is also called a sphere." "Let's see how far we can kick a ball." "I wonder how far I can kick this milk carton. Will it roll like a ball?" "I wonder why the carton doesn't go as far?"

Note:

Which sphere is the most fun? Balls, of course! Whenever you play with a sphere, mention the word and pretty soon the word will become a part of the children's vocabulary.

Wouldn't it be fun for a toddler to try to pick up a bowling ball? Compare a ping pong ball with a bowling ball.

Counting

Resource Books:

Vera Williams' *More, More, More Said the Baby*

Teri Sloat's *From One to One Hundred*—delightful counting book; a little old for toddler, but one of the few books with beginning math concepts that are available for young children

Nancy White Carlstrom's *Let's Count it Out, Jesse Bear*—learn how to add one more to each number, bouncy, fun rhymes, part of the Jesse Bear series

ToddlerTime™ *Numbers*

Activity:

Clap and/or march to the numbers one, two, three, and four—either music or voice. Toddlers love marching music.

To start a toddler with counting, chant the rhyme: One, two, Buckle my shoe.

Talk about counting, but don't count objects over the child's age—"One, two."

Rote counting* is OK because it will be a game for toddlers, but avoid drill and practice.

Sample Questions and Comments:

Rote Counting:* Put on the music or just start chanting. Remember that even if you can't sing you can chant. You won't need to say a thing. Toddlers will watch and then follow your lead.

Counting Objects: "How many crackers do you want? Two, one for each hand." "Lets count noses (Touch each nose in the group as they are counted.). That means we should make six sandwiches." "How many noses do you have? One nose. You have one nose and I have one nose. But you have two hands. One, two hands." "I see two balls." "There is only one trike, but we have two Big Wheels." "She has two teddy bear counters. I have one."

New Words:

Count/Counting	Chanting/Singing
Numeral Names	Names of Body Parts

Save for Older Children:

Symbols, Signs, and Numerals. *Step Ahead Golden Books Workbooks, Sticker Starters,* and *Mark and See* are a colorful, inexpensive series for preschoolers through Grade 3 and up. "Before I do Math" is appropriate.

*Note:

*Do not confuse rote counting with counting objects ("rational counting"). Repetition is one of toddlers' favorite pastimes, so rote counting comes fairly easily to them, although they will not YET understand that they are counting objects. They can begin to construct a mental picture of one to one relationships…one nose, two hands, two feet…these basics will lead to understanding counting later. Repeating counting rhymes and counting-in-order helps acquire the idea of sequencing.

*See also Math/Rote counting: Be sure you and/or the child touch whatever you count.

Fingerpaints

Materials:

Low table or easel, fingerpaints (with a little soap), paper, pail, sponges, etc.

Apron (if you can get one on the toddler)

Resource Books:

Kohl/Gainer's *Math Arts: Exploring Math Through Art for 3 to 6 Year Olds*

Vera Williams' *More, More, More Said the Baby*

Activity:

Set up the usual fingerpaints—be sure to roll up any long sleeves.

Ask toddler if she wants one or two spots of paint.

You can ostentatiously dole out one or two spots...counting out each spot of paint. Then offer a third spot. Talk about it. You can talk about two hands, two feet, or two ears. If you add the soap separately, you can also count out the dots of soap.

You probably already extend the exercise with talk about different colors, and what happens when several colors are mixed— the color of poop, of course.

New Words:

Count	Fingerpaint
Dab/Spot/Dollop/Bit	One/Two/Three
Add	More

Sample Questions or Comments:

"Would you like one or two spots of paint?" Count out, "One...two." "There, you have one dab for each hand (foot, elbow, ear...)." "Would you like one more spot of paint?" or "Shall we add one more spot of paint?" "Now, you have three spots of paint."

See also: Physics/Wet and Dry and Math/Counting

Note:

Toddlers love to play with the water and will be very happy to help you with cleanup. Keep lots of sponges nearby.

Following Instructions

You Will Need:

Everyday tasks to be done—e.g. clean up, set the table...

To Do:

Ask a toddler to perform one task. Then combine two tasks. Later try asking the toddler to complete a series of simple tasks.

Talk about "before," "after," "while," "during."

New Words:

job/task/assignment/chore
sponge/cloth/towel
clean/wash/wipe
hang up/attach/put

Sample Questions and Comments:

"Please put the paper into the trash can." "First will you take off your jacket, and then please hang the jacket on your peg." "Please go get the...?" "Please help us wash the table with a wet sponge, dry the table with a towel, and then you can put the napkins on the table for snack." "Thank you."

Save for Older Children:

Ostentatiously "reading the instructions" on anything. Older toddlers and preschoolers will love to help you follow the instructions, reading one step at a time.

Shoes

You Need:

Shoes and socks

SEE ALSO: Math/shoes

Resource Book:

Ray, R. *Whose shoes are these?*

New Words:

shoe   pair   sox   sneakers/sandal
two   match

To Do:

Talk about shoes. Compare them. There are dress shoes, tennis shoes, sandals, go-aheads, wedgies, slippers....

Have store catalogs with shoe pictures available. Ask children to find the ones that look the most like theirs.

Put everyone's shoes into a pile in the middle of a circle. Cover the pile with a cloth for a minute or two. Then lift the cloth and ask toddlers to go find the shoe that matches the one they have on. If there are too many children to let each child go individually, you can send the children in pairs, or make it a free-for-all for everyone to find his shoe at once. Hectic, but full of laughs.

Sample Questions and Comments:

"Let's each take off one shoe." "Now we will put that shoe in the middle of our circle." "Can you find the shoe that matches the one you have on?" "Now you have two shoes." "We have two hands and two eyes." "Are you two years old?"

Save for Older Children:

Competition to see who can find their shoe and put their shoes on first. Tying their shoes will come later too.

How Tall Am I?

You Will Need:

A place where you can record growth*

A straight edge (stored nearby)

Marker pen or pencil

Resource Book:

Moomaw/Hieronymous' *More than Counting: Whole Math Activities for Preschool & Kindergarten*

To Do:

Put the straight edge on a toddlers' head and mark the wall—this is a time-honored activity.

Be sure to mark regularly. The toddler will grow a lot between opening day and the last day of school.

Be sure to put the toddler's name and the date on each mark.

Measure the adults too.

New Words:

Tall   Measure   Big
Short
Small   Mark   Little
Marker

Sample Questions or Comments:

"I wonder how tall you are?" "Stand by the wall and I will make a mark to show how tall you are." "It looks like your Daddy is the tallest." "The Elephant is still taller than Daddy." "Babydoll is the smallest." "How many of you make one Daddy?" "How many pancakes big are you?"

Tip:

A young toddler will not know what is happening at first, but as she grows it will be fun to see how little she was.

Save for Older Children:

Assigning height numbers. Sharing time with clothes that the children have outgrown. Sharing baby pictures between your students.

***Note:**

Some schools have imaginative life-sized posters or paintings of large animals (an elephant, giraffe or even a Daddy) to use as measuring sites, some with inches and centimeters side by side. Sports stores carry life size cardboard standing cutouts and posters of famous sports figures. Children are impressed with the size comparison. "How many toddlers would have to stand on shoulders to equal Michael Jordan?"

Tape Measure

Things You Will Need:

A measuring tape, one for each child, if possible.

Resource Books:

McBratney's *Guess How Much I Love You*

Aliki's *I'm Growing,* Stage l, Let's Read and Find-out Science series.

What to Do:

Let each toddler manipulate the tape measure.

Measure things and talk about how big or little things are.

Measure each toddler. Measure yourself. Measure Mr. Bear.

New Words:

measure	size	tape
tall	little	bigger/smaller

Sample Questions or Comments:

"How tall are you? How tall am I?" "Mr. Bear is this tall. It takes two bears to make one of you. You are twice as tall as Mr. Bear." "You are this tall. It takes two of your measure to make one of me."

"I wonder if that table will fit into this corner. Let's measure it." "Lets see how long this block is." "Is this block longer or shorter than that one?" "Will you hold that end of the measuring tape?"

"How many steps does it take you to go between the table and the wall?" "Let's see how many steps it takes me to go between the table and the wall. It takes you seven steps and it takes Teacher three steps." "My legs are longer than yours."

Tip:

If left unsupervised, toddlers can easily ruin an adult measuring tape. There is a tape on the market called THE BIG TAPE™ made with children in mind.

Save for Older Children:

Numbers, comparisons, and predicting how long an item is. Talking about inches and meters.

Note:

A toddler will be fascinated with the measuring tape and could probably be happy to simply play with it. Give that toddler time. You will be surprised at what a toddler will think to do with it. Then you can talk about it.

Dancing

You Need:

Cassettes or CDs with rhythmic music—marches are always good

Plenty of room to "dance" or march.

—or—

Hands for clapping and a voice for chanting

Try: Jane Marino & Dorothy F. Houlihan's *Mother Goose Time: Programs for Babies and Their Caregivers*.

Rediscover the rhythms of language through traditional rhymes, songs and finger plays, including such favorites as "Head Shoulders Knees and Toes," "Clap, Clap, Clap your Hands" and "Five Little Monkeys."

To Do:

Turn on the music and dance, hum, swing your hips, tap your feet, clap your hands. Clap your hands to different beats. Alternate one clap with two claps and repeat the pattern so that the children follow you. Stamp your feet in a pattern. You don't even need music to do this.

Resource Book: Helen Oxenbury's *Clap Hands*

New Words:

music/march, rock, country	dance
clap, stamp, wiggle, tap	hum

Sample Comments & Questions:

"This is my favorite marching music." "Follow Jason." "Can you hum?" "Let's clap our hands" (try different rhythms—clap twice, clap twice fast, clap three times slowly—count as you do so).

LATER: Everybody do the Hokey Pokey, or other circle songs (Moms and teachers too), great fun.

Learning to take instructions, identifying body parts, and just having fun in a crowd.

Note:

Plato once said that "music is a most potent instrument than any other for education." At no age is that more true than during toddlerhood. Music, scientists believe, trains the brain for higher forms of thinking. Turn on the music and watch your toddler enjoy!

Save for Older Children:

Action songs that require following instructions will be more fun later. However, toddlers will enjoy watching or following older children on a march. They love to follow the leader, and can shake wrist bells or any one piece noise maker.

Tip:

You can use music to accomplish many things, e.g. "Drop my toys, pick em up, pick em up" (as you pick up toys), "Wiggle the giggle right out of my self," "Goodbyyyyeee, goodbyyyyeeee, see you tomorrow."

Playdough Math

You Need:

Playdough

plastic knives (introduce later)

To Do:

Give each toddler a piece of dough and let them squish, roll and knead it. Talk about each person is getting one piece and how many pieces will be needed. Then cut one of their pieces in half. Now each toddler has two pieces. Distribute plastic knives and show toddlers how to cut.* We live in a world of numbers. Toddlers can start "beholding" ideas, but not numerals yet.

You can tear a piece of paper in half and get two pieces—talk about it.

Good Resource Books:

Sara Atherlay's *Math in the Bath (and other places too)*

Nancy White Carlstrom's *Let's Count it Out, Jesse Bear*

Usborne Books at Home, Board books for babies and toddlers: *Playdough*

New Words:

dough	knife/plastic
knead/squish/roll	count
pieces	cut
each	person

Sample Questions and Comments:

"How many pieces of dough do we need?" "Let's count noses." "How many noses do you have Jesse?" "OK, one, two, three four noses." "So I need four pieces of dough." Now, we each have one piece of dough." Would you like to have two piece of dough?" "I will cut the dough in half." "There, now you have two pieces—one, two." "One two, Buckle my shoe. Three four, shut the door."

Save for Older Children:

Talking about fractions, ratios and scale.

*Note:

Young children tend to press down on a knife and need to be shown how to "saw" with a knife.

Opposites

You Need:

Any opportunity to talk about opposites:

BIG/LITTLE	WORK/PLAY
REAL/PRETEND	MESSY/NEAT
LIGHT/DARK	COLD/HOT/WARM/COOL
HEAVY/LIGHT	IN/OUT
SHORT/TALL	BOTTOM/TOP
HAPPY/SAD	INSIDE/OUTSIDE
HARD/SOFT	ON/OFF
WET/DRY	

Resource Book:

Rainbow Books, *Opposites*—colorful, inexpensive board book with lively photographs

Fit-A-Shape™ *Opposites*   *Dr. Seuss foot Book*

To Do:

Take advantage of opportunities to identify and talk about opposites—set the stage if necessary. Throughout the book you will find ideas.

In cooking: Take note of wet and dry ingredients.

In the sandbox: Note that wet sand pours, dryer sand packs and molds, very dry sand blows and drifts like snow.

Grownups and their belongings are big and toddlers and their belongings are small. A real dog is big and a plastic one is little.

New Words:

See list at left, the concepts are endless opposite/opposites

Sample Comments or Questions:

"Kevin isn't cold, he is just the opposite—he is hot." "The flour is dry and powdery. The milk is wet. Let's mix the dry ingredients with the wet ingredients." "The elephant is big. Your teddy bear is little." "But you are bigger than Mr. Bear."

Save for Older Children:

Playing the "Opposites" guessing games.

Shoes

You Will Need:

Shoes: little shoes, big shoes, sandals, boots…

Socks

—and/or—

Pictures of all kinds of shoes from store catalogs and books

Resource Books:

R. Ray's *Whose Shoes Are These?*

Istar Schwager's *Active Minds*—fine colorful photos of tennis shoes, fingers, etc.

See also: Math/matching shoes

To Do:

Whenever shoes are being taken off or put on, seize the opportunity to talk about differences and similarities and/or the number two—a pair. Try to include adult's shoes in the mix. During the summer there are pairs of sandals & tennis shoes. In the winter there are pairs of boots—rubber boots, hiking boots, fireman's boots, astronaut's boots and ski boots.

You can find books that have pictures of different shoes and talk about them. Store catalogs are especially complete.

Questions and Comments:

"I wonder whose shoe this is?" "Now you have two shoes." "Jason's shoes are just like Kevin's shoes." "I wonder if your shoe would fit me?" "Your Daddy's shoe is big." "You have two shoes, two socks. You have a pair of six and a pair of shoes." "…and you are two years old."

New Words:

shoe, socks, match, pair, two, big and little, sandals, boot, fuzzy slippers, the colors (if you want to get into that).

Save for Older Children:

✔ Talking about dress shoes and sport shoes—Does Teacher wear the same shoes as Michael Jordan? See also Math/matching shoes.

✔ Egyptians wore sandals. In what country did they wear those wonderful curly toed-shoes?

Talking about pairs—sets & twins. Then, pears for a little humor. Can you pare a pair of pears? Four year olds will think that is funny.

Pegs and Pegboards

You Will Need:

The old favorite Pegs and a few pegboards. There are a number of different companies that offer them. We like the pegs that are shaped like a handle that fit into small toddler hands—called easy-grip. Punkinhead pegs™ are also easy to manage.

To Do:

Place peg boards on the table with a few pegs in the boards as a model. It is not necessary for toddlers to sit down.

Keep the supply of pegs up away from toddler unless you are including dumping as a part of your project.

Ask a toddler how many pegs he wants. If he cannot answer, put one peg into each hand. You can continue to add more pegs as the project continues, counting them as you add.

Enjoy a peg board yourself. The toddler will watch. Make a pattern: red/blue/red/blue and talk about it. Enjoy what you do and share the joy. Take the opportunity to talk about simple numbers and colors.

Don't be afraid to use the word "manipulative." Some nursery schools have manipulative areas available daily.

New Words:

peg   blue/red/yellow   "how many"
pegboard   manipulatives
one/two-handful

Sample Questions and Comments:

"How many pegs would you like?" "Here are two pegs, one for each hand." "Look, you have one red peg and one blue peg." "Can you fit the pegs into the holes?" "Lets pick up

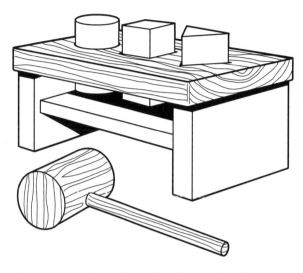

the pegs and put them back into the basket." "You have a handful of pegs." "Can you dump the pegs into this container?" "Let's put all the blue pegs in this bowl."

Save for Older Children:

Manipulating, sorting, patterning, counting, or wrapping string or rubber bands around the pegs.

Note:

The toddler will not fully absorb number concepts yet, but pegs and mats are a great start for him. Counting songs and chants will bolster math learning. Don't be surprised if toddler dumps the container of pegs with a lovely crash. Pegs, like bear counters are fun to just gather up and dribble onto the table, into a bowl or onto the floor, over and over again. You might provide two larger empty containers in which to dump the pegs back and forth.

Rote Counting

Resource Book:

Helen Oxenbery's Clap Hands

Musical Resource:

Marino & Houlihan, *Mother Goose Time: Programs for Babies and Their Caregivers.* Includes favorites like "Clap, Clap, Clap Your Hands" and "Five Little Monkeys." See Rhymes and Chants.

Activity:

Put on some marching music or just chant a rhyme while clapping your hands. March together or play follow-the-leader. While sitting, you can still clap your hands and chant.

Use rhymes like: "One, two, buckle my shoe. Three, four, shut the door. Five, six, pick up sticks;" "One for the money, Two for the show, Three to make ready, And four to go;" or "One potato, two potato, three potato, four; Five potato, six potato, seven potato, more." (See appendix for more rhymes). Even a young toddler will catch the spirit and the patterns. Rhymes expose children to the music that is contained in our language.

Questions and Comments:

"Where are your hands? Let's clap our hands. Clap, clap, clap your hands."

"How many sandwiches do we need? Let's count noses. (Count each nose in the group.) How many noses do you have? One nose. You have one nose and I have one nose, but you have two hands. One, two hands."

New Words:

Chant/March   Count   Numeral Names
Names of Body Parts

Follow-the-Leader   Clap, Stamp, Hop

Save for Older Children:

Symbols, Signs, and Numerals.

*See the Appendix for an informal description of what to expect during the Ages & Stages of toddlerhood.

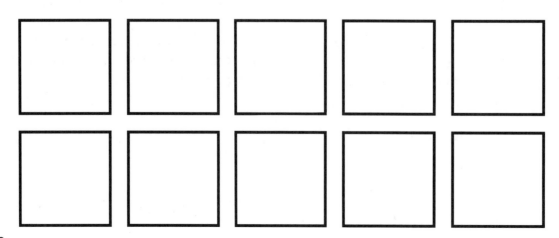

Dipping in Sequence

You Will Need:

Strawberries, crackers, green pepper slices, peeled cucumber "coins" or other ready to eat food that can be dipped

Two dips—one wet & one dry (sour cream, thinned cottage cheese (a toddler staple), sugar, whipped cream, mayonnaise…Jello powder is interesting) in small dishes. It is nice for each toddler to have two individual dipping dishes, but sharing is a learning experience too.

Tip:

Very young toddlers will want to empty the dishes—pouring is a favorite activity.

See also: Math/following instructions

To Do:

Serve food with two dips available. Tell toddlers to dip the food into one dip and then into the other, wet first, then dry. Talk about first dipping into one sauce and then, next, dipping into the other sauce. Talk about how first we do this and then we do that. Talk about other sequences and other things that come in twos (ears, hands*).

New Words:

Strawberries/cucumber	crackers
cream/mayonnaise	dip
sugar	wet/dry

Sample Questions or Comments:

"Can you guess what we have in this bowl number one?" "Lets taste it. It looks like salt. No, I don't think salt is that sweet. Of course, this is sugar. What do you think is in bowl number two?" "I think you are right—milk." "Now, first we'll dip this strawberry into the milk to get it wet, and then we'll dip it into the sugar." "I'll bet it will taste good."

Fun for Older Children:

Carrots, celery, cauliflower pieces & broccoli are all favorites for older children. Having mathematical snack times—sorting and classifying edibles. Using Goldfish™ popcorn or oyster crackers as counters.

*Note:

If humans had come fitted out with four fingers on each hand, then we would have undoubtedly developed a number system on a base eight rather than base ten. Our decades would be octades. We wouldn't talk about the Gay Nineties or the Roaring Twenties. Since we have ten fingers and ten toes a toddler will learn our number system on base ten. As we have said, learning math should be personally relevant.

Sorting and Categorizing

You Will Need:

Anything that is ready to be put-where-it-belongs: groceries, blocks, clean dishes, clean clothes or toys

The old standby shape sorting cubes or buckets

Useful Books:

Sara Atherlay's *Math in the Bath (And Other Places Too)*

Toddler Time™, *Sorting,* McClanahan Book Co.: inexpensive, bright and full of good ideas—the best part is the "Toddler Tip" to help integrate the skills taught into parents' daily routines

To Do:

Let him put away the spoons

Let toddlers help you put things where they go. They can put the spoons in the spoon tray, canned goods in the cupboard, or round blocks with the other round blocks. They will be categorizing, classifying and sorting—all good basic math concepts. Don't forget to talk about it.

The commercially available shape sorting cubes or buckets ought to be an integral part of your manipulative corner and are most

useful when words are put to actions.

New Words:

Names of items being put away—e.g. spoons, cans

Names of the categories of the items—e.g. groceries

Sort, kind, category, shape

Sample Questions and Comments:

"Let's empty the dishwasher. Will you sort out the spoons, and put all the spoons in this drawer with the other spoons? Thanks." "We have a lot of groceries to put away. You can put all the cans on the can shelf." "Let's put away these blocks. You can put away all the round blocks. They go here with the other round blocks." "The bear counters go into this basket."

Save for Older Children:

Sorting, categorizing and labeling collections of things—shells, rocks, baseball cards…

See also: Geometry

Note:

A great rainy-day activity: go through a jumbled container of toys and organize them. Put all the stray pegs into their basket, all the puzzle pieces into a basket, the blocks in the block corner, the trash in the trash can, and so on. We used to dump everything on a big blanket and dig in. The children discovered things they hadn't seen in months. Items were categorized and put-where-they belonged. What was left on the blanket got tossed in the Goodwill box.

Timer

You Will Need:

A wind-up, ticking kitchen timer

To Do:

Bring out the new timer and tell toddlers what it does, e.g. it ticks, it rings when it is time to eat snack, etc. Use the timer when it is time for transition. Somehow the timer is neutral and nonjudgemental. It simply tells us that it is time to do something else.

Let toddlers carefully handle the timer and show them how it winds.

New Words:

when   wait   timer   ring   tick/ticking

Sample Questions and Comments:

"Can you hear what the timer is saying?" "It sounds like it is saying 'tick tick.'" "When the timer rings it will be time for snack." "When the timer rings it will be time for naps." "We will wait until the timer rings." "Waiting is hard to do." "Let's set the timer so it can tell us when it is time to go home."

Save for Older Children:

Timeouts, timed competitions, and telling time.

Note:

It is fun to have a collection of timers and clocks, and compare the different rings and ticks.

Notes

Nutrition & Health

"Studies show that variety and good nutrition go hand in hand."

Journal of the American Dietetic Association, *March '97*

My Objectives

Nutrition

As you use the following Nutrition Activity Pages keep in mind the secrets of successful and healthy snacks:

Snacks are important to toddlers—they have small tummies.

Let the kids be a part of food preparation.

Teach children that food helps them grow, learn and play.

Food can be fun—it is a part of social life.

Variety is the spice of life.

We cannot teach toddlers about good nutrition, but we can be good models and help toddlers begin to develop healthy eating habits. Helping to prepare food and eating with others helps toddlers learn by example. When children are older they will understand the whole sequence of picking or buying, bringing home, fixing, eating food and doing the dishes, and will want to help with it all. As you cook with toddlers, however, you will want to do most of the food preparation and set-up by yourself ahead of time so that toddlers can "just do it."

Toddler Tips:

Young toddlers who are just shifting from breast milk or formula need extra fat to provide enough energy for brain development and growth. After age 2 you can gradually lower the fat in a child's diet. Because of the calcium requirement a toddler needs at least two full servings (1/2c) of milk or milk prod-ucts each day. Yogurt is the best source of calcium, with nearly 500 milligrams of calcium per cup. A cup of milk contains about 300 mg of calcium.

Because a toddler's tummy is small, serve about 1/4 of an adult serving. Variety should take priority.

Good food is good no matter when it is eaten. "Hors d'oeuvres" of fruit or vegetables won't spoil toddler's appetite and are a good way to accumulate the five or more servings recommended. Sugar or fat will spoil the appetite, however.

Toddlers like to serve themselves and drink from a cup. Try nice warm soup on a cold day. It warms the nose as well as the tummy. When possible, little plastic measuring cups with a pouring lip make fine individual serving pitchers. The adult pours from the large container.

A Toddler Cooks and Learns:

Toddlers are exposed to physics, chemistry, math and nutrition as they watch and help cook. They are good at washing hands, wiping counters and table tops, helping harvest in the garden, helping cook, breaking, scrubbing, snapping, tearing, stirring, chucking corn, and dipping food items.

When cooking with toddlers choose flexible recipes that are difficult to mess up.

New Cooking Words:

The list is endless:

Knead, melt, slice, sift, stir, break, scrub, snap, tear, chuck corn, hot, cold, tasty, sour, sweet and dip

Good Resource Books:

Chubby English/Spanish Books, *Let's Eat; Vamos a Comer.*

Lois Ehlert's *Eating the Alphabet; Fruits and Vegetables from A to Z*

Patricia Barrett-Dragan and Rosemary Dalton's *The Kid's Cookbook; Yum! I Eat it. Good*

Katzen & Henderson's *Pretend Soup & Other Real Recipes: A Cookbook for Preschoolers & Up*

New England Dairy and Food Council. *Nutritious Books for Young Readers Pamphlet,* Boston. A free list of books that are widely available at bookstores. They make learning about nutrition fun and interesting.

Note: Some foods that don't cause cavities: eggs, cheeses, ham, peanuts or walnuts.

Teacher Notes:

_____

_____

_____

_____

_____

_____

_____

_____

_____

_____

_____

_____

_____

_____

_____

Trying New Foods

You Will Need:

An interesting new food—try one food or one combination at a time. It is fun to have children share an unusual food that they like—artichoke, strawberry, mango, coconut…

Fun books to read: Sarah Hayes' *Eat up, Gemma*, Mitchell Sarmat's *Gregory, the Terrible Eater,* and Mary Ann Hoberman's *Seven Silly Eaters*.

Resource Book:

Gregory, Valiska, *Riddle Soup*, NY. Macmillan Inc. 1987: shows parents how guessing games like riddles can reinforce nutrition learning and stimulate children's interest in trying new foods

Musical Resource:

Zucchini Brothers' *In Your Garden,* includes "Eat all Kinds of Things"

To Do:

Wait until toddlers are hungry. They will be more adventuresome when others are also tasting a new food.

Set the stage. Offer food that is easy to eat—one bite cracker, cut banana (leave skin on so toddler can peel the slice), squares of cereal (Ralston™ wheat chex are healthy), orange "smiles" (cut off the white), strawberry (cut in half), grapes (cut in half). A pretzel can be used as a spear. Experiment with florets of broccoli or cauliflower ("baby trees") dipped in thinned creme cheese or artichokes in melted butter. Compare different kinds of cheese. Cheese comes in slices, in chunks, dips or strings (some toddlers love it).

Toddlers like to serve themselves. Little plastic measuring cups make serving pitchers. The adult pours from the large pitcher.

If the first introduction of a food doesn't work, wait, and try again under better circumstances.

New Words:

name of the food	grape	measure
broccoli	dip	mix cheese
artichoke	bite	description of the

food—crispy, sweet, sour, soft...

Sample Questions or Comments:

Talk about what you are preparing and how it is eaten. Ask, "What do you think will happen when…?" Discuss the color, the texture, and the smell. Sometimes, while eating, it is smart to keep comments—other than "yum," to a minimum. Toddlers may interpret any comments as pressure to try new things. Taster Rule: if you like the food, say so. If you don't, zip up your lip.

TIP: Protein is important for growth. It is good to know that a peanut butter sandwich (with wheat bread), macaroni and cheese, a cheese sandwich, or rice with black-eyed peas are all complete protein combinations. A glass of milk added to these makes a healthy meal and, best of all, most children like these simple combinations.

ANOTHER TIP: When introducing yogurt, try serving it in an ice cream cone…and not necessarily frozen yogurt.

Save for Older Children:

Trying exotic spicy foods. Reading and then making "Stone Soup" (the folk tale).

Apple Circles

You Will Need:

Apples (one apple will serve about three people)

apple corer

popsicle sticks

peanut butter (without sugar)*

knife (for adult)

Resource Book:

Lois Ehlert's *Eating the Alphabet: Fruits and Vegetables from A to Z*

To Do:

Roll up cooks' long sleeves.

Wash the apples. Toddlers will enjoy this. Core the apples, removing the stem and seeds.

Talk about seeds to older toddlers. Stuff the opening with peanut butter. To serve, cut crosswise.

"Look! You cut a circle!" Don't forget to describe the sequence—"first we…, now you can… and then we will…"

New Words:

apple   green/red   peanut butter
stuff/pack/cram   slice
wheel/circle   middle   core

Sample Questions or Comments:

"Look at this. Tell me what you see." "Are you sure it is not an orange?" "No, you are right, it is an apple." "Does anyone know what this is?" "Taste it and see if you can guess—of course, it is our old favorite, peanut butter." "We are going to make apple pinwheels. First we will wash the apples. Then I will cut out the seeds in the middle of the apples. See, it leaves a hole." "Now we can put some peanut butter into the hole." "Here is a stick for you to use to stuff the peanut butter into the hole." "How do you suppose we can make an circle out of this apple?" "That is a good guess. Let's try this…"

Tip:

The apple/peanut butter slice is smoother if it has been chilled before cutting, but the continuity is apt to be lost on toddlers.

Save for Older Children:

Presenting an array of apples of different colors and kinds, and tasting and comparing them. There are 7,500 varieties of apples throughout the world.

*Note:

Cream cheese or cheese spread can also be used. It it important to chill cheese before slicing.

Fact:

The apple tree is a member of the rose family.

Beatin' Batter Cookies

You Will Need:

Makes two dozen

4 1/2 c. Oatmeal

2 1/4 c. butter or margarine

1 c. brown sugar

1 1/4 tsp, baking powder

2 c. flour

1/4 c. applesauce

Big bowl, a few spoons, and measuring cups. Two forks (salad forks are a good size and shape for toddlers.)

To Do:

Roll up long sleeves and wash cooks' hands.*

Preheat oven to 350° and put all utensils and equipment on a table so toddlers can see them. Talk about them. Keep ingredients up where toddlers cannot get them until it is time to add each item. Adults do the measuring—into more than one measuring cup. Let toddlers help dump all the ingredients into a large bowl. Stir it, mash it, pound it, and knead it. The longer and harder you mix it, the better it tastes. Each toddler can have a bowl. It won't hurt to taste it. Show toddler how to roll the dough into small balls. Bake on cookie sheet for 10 to 12 minutes. TALK ABOUT ALL OF IT—measuring, the sequence of events, mixing wet and dry ingredients. While the cookies are baking enjoy the smells as you clean up the area* and the toddlers. Then as you eat, read stories about the gingerbread man or Cookie Monster.

New Words:

spoon	measuring cup	oatmeal
brown sugar	flour	butter
baking powder	mash	
knead/pound		

Sample Questions or Comments:

"These are the things we need to bake cookies." "We have enough spoons so three people can stir. One, two, three spoons." "This is a measuring cup. We have two cups so two people can dump ingredients at the same time." "First we put in the margarine—it looks like butter doesn't it?" "We will use forks to smash the butter."  "Now it is time to put in the dry ingredients. The oatmeal is dry and so is the brown sugar and the baking powder." "Look how white the baking powder is. It looks like the flour." "Look, the little balls we made turned into circles."

*Note:

Plan plenty of time for these tasks. Toddlers love to wash—anything.

Crackers

You Will Need:

1 package oyster crackers (avoid brands with palm oil or lard)

2 tablespoons canola or other vegetable oil

1 salad dressing packet—dry (The favorite is Ranch dressing mix, but others can be tried). Parmesan cheese may be substituted.

Good Reference Book:

Smith, Kathie Billingslea, and Victor Crenson, *Tasting*

To Do:

Place crackers into a large zip plastic bag. Shake the crackers as you sprinkle oil over them. Zip bag and shake gently until crackers are coated. Pour coated crackers onto paper towels and let stand for 10 minutes. Return to bag (or to individual bags) and sprinkle dry dressing or cheese over crackers. Zip bag(s) again and shake to mix well. The crackers are now ready to eat. Pour into a bowl or eat from the bag.

New Words:

oyster crackers	shake	bowl
oil	flavoring	
paper towel	salad dressing	
parmesan cheese	plastic bag	salty

Sample Questions or Comments:

"The food is salty." "The crackers are crispy." (When trying new foods it is smart to keep judgmental comments to a minimum.)

Note:

Follow good taster rules. If you like the new food, say so. If you don't zip up your lip.

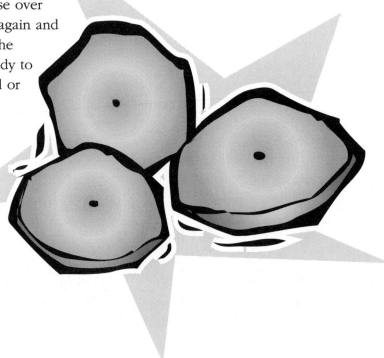

Fruit Salad

You Will Need:

7 Servings

   1 small plain yogurt

   3 bananas

   2 apples

   2 tangerines or oranges

   1/4 c orange juice

   other in season fruit (e.g. grapes*)

   bowl, spoons, serving cups

You can ask the children to bring a fruit to share.

Useful Resource Books:

   Barbara Lember's *A Book of Fruit,* NY, Ticknor & Fields Books for Young Readers

   Gillimard Jeunesse and Pascale de Bourgoing's *Fruit,* Scholastic First Discovery Books

   Lois Ehlert's *Eating the Alphabet: Fruits and Vegetables from A to Z*

   Snapshot™ Books, *Things That Grow*

To Do:

   Wash cooks' hands and roll up long sleeves.

   Let toddlers examine, discuss and watch the fruit being prepared and help put it into a bowl. Older toddlers can cut bananas. (Grapes should be cut in half to prevent choking.) Mix yogurt and orange juice. A toddler can pour mixture over the fruit. Toddlers can stir. Chill until snack time. Toddlers can also break, scrub, snap, and tear foods.

Idea:

   A can of crushed pineapple or mixed fruit with the juice can be added to keep the bananas from browning.

New Words:

Peach   apple   tangerine/orange
orange juice   Yogurt
other fruit names

Sample Questions or Comments:

   "What is your favorite fruit?" "Thank you for bringing a banana, Josh. I will help you cut it." "I wonder what will happen when we put the orange juice into the white yogurt." "Would you like to pour the mixture over the fruit?" "Now it is time to stir. Here is a big spoon."

Nutrition Tip:

   Peaches, apricots, cantaloupes, bananas, nectarines, mangos, and watermelon are sources of carotene. Grapefruits, oranges and other citrus and juices, melons, berries, papayas, and kiwi are all good sources of vitamin C, among other things.

   Fruits provide vitamins, minerals and fiber and have no cholesterol and little or no fat or sodium.

Save for Older Children:

   Cutting hard fruit by themselves. Following a recipe.

Graham-Apple Sandwich

You Will Need:

Sweetened applesauce

graham crackers*

Whipped cream (optional)

Plastic knife for each toddler

serving spoon

Resource Book:

Anne Christenberry and Barbara Steven's *Can Piaget Cook?*

To Do:

Give each toddler a graham cracker. Put a spoon full of apple sauce on a graham cracker. Ask toddler to first spread the applesauce well on the cracker, then put the second cracker on the top. Talk about the sequence: first, then, and then. Whipped cream can be added.

New Words:

graham cracker

applesauce

sandwich

spread

on top of, underneath

first/then

knife

crunch

Sample Questions or Comments:

"Can you guess what this is?" "Right, a graham cracker." "Taste this and see if you can guess what it is—yes, applesauce." "We are going to make a graham/apple sandwich." "First you spread the applesauce on the cracker." "Then we will put a cracker on top—now we have a sandwich." "Then we can put a dab of whipped cream."

Tip:

You can add a third cracker and make a triple decker sandwich. Then you can talk about three crackers.

See also: Math/sequences

*Note:

Graham flour, the basis for graham crackers is a more coarsely ground whole-kernel flour than ordinary whole wheat, both vastly more nourishing than white refined flour. Animal crackers, gingersnaps and fig bars are also low in fat and sugar.

"Party" Mix

You Need:

4 cups crunchy cereal (e.g.Cheerios®, Ralston Wheat Chex® or any combination)

1 cup pretzel sticks (smallest size)

1 cup seasoned croutons (optional)

No oil, seasonings or baking is necessary for toddlers.

To Do:

Put each item in an individual bowl. Give the toddler a cup and show him how to put a little of each item into his cup. Give him a second cup and show him how to pour the items from one cup to the other, mixing them. There will be some spills so have the toddler pour his ingredients over a cloth so the spills can be put back into the cup, or do this activity outdoors so the birds can help with cleanup.

New Words:

dry cereal   mix   choose
little bit   pretzel   spill
crunchy

Nutrition Tips:

It is fortunate that the favorite foods of most young children—bread, cereal, rice and pasta, are at the important base of the Food Guide Pyramid with a recommendation of 6–11 small servings per day.

See also: Nutrition/grains and Nutrition/crackers

Feed toddlers plenty of fiber. To figure out how many grams of fiber a child needs, add five to his age (2+5=7 grams). Other good fiber sources also include, apricots, prunes, small broccoli & cauliflower ("baby trees"), unpeeled apples (one apple has 3 grams).

Also encourage toddlers to drink more water—much better than so-called fruit drinks that are mostly sugar, corn syrup and food coloring. Triskets, also made of whole wheat, are a good substitute for mostly-white-flour crackers.

Save for Older Children:

Following the recipe for "party mix" on the back of a cereal box.

Pop! Pop! Popcorn!

You Need:

Popping corn

Pot for popping—electric pan (so the children can see and hear the process)

Oil (canola is good) just enough to coat the pan

salt and butter (though not necessary)

—*and/or*—

Corn cereals, corn on the cob, (popcorn on the cob), corn bread, hominy*

(Incidentally Sugar Corn Pops are almost half sugar).

To Do:

It is delightful to watch corn pop outside, without a lid. (Birds will eat it.) Older toddlers will enjoy pretending to be popcorn. It is fun to have "oil" poured over you as you quietly wait to get "hot" enough to POP. Even popping corn comes from a cob. Popped corn can float in soup, can be coated with brown sugar and cinnamon, can be fed to the birds, or put into collages, strung as decorations, or be even used as "counters." At snack time it is interesting to compare corn products.

Popcorn song (tune: Row, Row, Row Your Boat):

"Pop, pop, pop your corn (or my corn). Pop it big and white…"

Safety Tip:

Wait until toddlers are off the bottle before you give them popcorn—very little ones can choke on popcorn.

Nutrition Tips:

It is fortunate that the favorite foods of most young children—bread, cereal, rice and pasta, are at the important base of the Food Guide Pyramid with a recommendation of 6–11 small servings per day. Cheerios®, the whole grain oat cereal, contains about twice as much protein as corn flakes, and most grains, wheat especially, are a more nutritious than corn, but corn is a grain, and presents all of grain's attributes in the most interesting variety of forms to small children. A diet that is low in fiber or fluid can cause constipation—a tough problem when it happens to a toddler. Feed toddlers plenty of fiber—to figure out how many grams of fiber a child needs, add five to her age (2+5=7 grams). Good fiber sources also include, apricots, prunes, small broccoli & cauliflower ("baby trees"), whole-grain breads, unpeeled apples (one apple has 3 grams) and fiber-enriched cereals—Ralston Wheat Chex® is a good source (3/4 c raisin bran has 5 grams). Also encourage toddlers to drink more water—much better than so called fruit drinks that are mostly sugar, corn syrup and food coloring. Triskets®, made of whole wheat, are a good substitute for mostly-white-flour crackers.

*Note:

Toddlers will enjoy experimenting with whatever forms of corn products they are given, but that "aha" moment that affirms that they belong to a category called "corn" will not come until they are older.

Grandma's Soup

(A Good Thanksgiving Activity)

You Will Need:

Vegetables brought from toddlers' homes (ask parents ahead)

3/4 c. onions, chopped

2 beef bouillon cubes

water

Large very clean stone

Large pot (a crockpot is fun to watch if enough adults are close by)

A large dishpan for toddlers to wash vegetables. Brushes are fun to have.

Resource Books:

Smith, Kathie Billingslea, and Victor Crenson, *Tasting*

Katzen, Mollie & Ann Henderson, *Pretend Soup and Other Real Recipes: A Cookbook for Preschoolers & Up*

To Do:

("Grandma is old and needs help to make her soup…" make up a story emphasizing the magic of sharing—a takeoff of the classic Stone Soup.)

Roll up long sleeves and wash cooks' hands.

Clean a toddler height table. Toddlers love to wash the table and the veggies. Let toddlers watch as the donated vegetables are pared and chopped. Cut veggies so that they will cook soon.

Saute onions before adding the other ingredients. If you like, you can save out celery, frozen peas and other fast cooking items until the last few minutes. Savor the lovely smells and serve for snack with crackers. If you have a lot to feed, a package of soup mix can be discreetly added.

New Words:

vegetable names	wash/scrub/clean	
brown onions	peel/cut	hot

Sample Questions or Comments:

"Thank you for bringing the carrot, Brian." "Let's cut the vegetables small so they will cook faster." "Is it soup yet?" "What vegetable is that? It looks different now that it is cooked." Sometimes it is smart to keep comments other than "Doesn't it smell good!" to a minimum.

Note:

Toddlers won't understand the classic story of Stone Soup, or even the toddler version—Grandma's soup, but will enjoy the fact that they have contributed something. Be sure to credit who brought what. On a cold day in November this might be a successful introduction of soup to toddler.

Save for Older Children:

Reading and enacting the classic story of Stone Soup…another great Thanksgiving activity.

Pancakes & Circles

You Will Need:

Package of pancake mix and milk or water
(see recipe on box)

—or—

2 C flour	1 Tbl. sugar
4 tsp. baking powder	1/2 tsp. salt

2 jars strained baby carrots or apple sauce

1 1/2 C milk (batter to consistency of thick
cream)

2 eggs	2 Tbl. oil

Large Bowl, Large spoon or two

Griddle or electric fry pan, spatula

A little oil

Inspiration

Butter, syrup or jam (optional) and forks,
plates (paper)

Foods that are or can be sliced to make circles e.g. slices of orange, apple, banana or
pear.

To Do:

Have cooks wash hands and roll up long
sleeves.

Follow instructions on the package, or mix
dry ingredients in one bowl and wet in
another. Talk about how much you will need
for each person. Ceremoniously measure
each cup. Let a toddler pour from the measuring cup by himself and then stir. Talk
about adding wet ingredients to dry. Talk
about the sequence as you read the
recipe. Toddlers like to stir.
Ask the children what
shapes they would like
you to make in the pan.
Suggest circle, triangle, "s"
and talk about shapes.
Pancakes are ready to turn

when they are full of bubbles. Extend the
activity by eating other food circles.

Useful Resource Book:

Feldman, Judy, *Shapes in Nature*, Chicago.
Eyewitness Juniors

New Words:

pancake	milk or water	hot
browned	spoon, bowl, cup	griddle
spatula	mix	
shape: triangle, circle, "s"		
ingredient	flip	wet/dry
squiggle	stir	

Sample Questions or Comments:

"Lets make pancakes." "How much shall
we make? How many people will want pancakes?" "Here is the bowl. I will measure
three cups of pancake mix and three toddlers
can pour the dry flour into the bowl." "What
do you suppose will happen to the flour
when we add this milk?" "The milk is wet,
isn't it?" "You can stir." "Shall we make a circle or triangle shape?" "Circles are easier
aren't they!" "I see little bubbles popped on
the pancake. It is time to flip it."

See also: Math: Geometry/Circle fun,
Math/Pancake circles, and Geometry/More
Circles

Save for Older Children:

Making faces with
raisins etc. Talking about
recipes, temperatures, and
vitamin content of pancakes.
Making and cooking pan-
cakes by themselves.

Pizza

You Will Need:

(10 servings)

   1 tube biscuits

   6 oz. mozzarella or cheddar cheese

   4 oz jar pizza sauce

   grater, wax paper, shaker for flour (optional), cookie sheet or foil, spoons

To Do:

Wash cooks' hands. Roll up long sleeves.

Let toddlers help you clean a low table. Let toddlers enjoy sprinkling the table with flour.

Have each toddler press out at least one biscuit. Rollers can be used if you have them.

Give each toddler a spoonful of pizza sauce to spread on the biscuit (or more if necessary). Adults will grate cheese while toddlers watch and talk about it. Let each toddler sprinkle cheese on her pizza with her fingers. Mark each pizza with a strip of paper. Bake at 350° until cheese is melted (10-15 minutes).

Use waxed paper under the cheese sprinkling so the spilled cheese can be recycled.

Resource Book:

Christenberry, Anne & Barbara Stevens, *Can Piaget Cook?*

New Words:

pizza	cheese	biscuits
pizza sauce	hot	roll, make flat
melt	sprinkle	bake, brown

Sample Question or Comments:

Sometimes it is smart to keep comments other than "yum" or "Doesn't it smell good!" to a minimum. Toddlers seem to interpret comments as pressure to try new things.

Note:

Try to allow toddlers to watch and smell the pizzas cook. Talk about the cheese melting.

Notice that pizzas are circles.

English muffins or french bread can be substituted, but toddlers enjoy kneading and flattening the dough.

Save for Older Children:

Making pizza from scratch and grating the cheese—veggie pizza is delicious, and discussing the nutritional value of pizza—proteins, vitamins etc.

Quesadillas

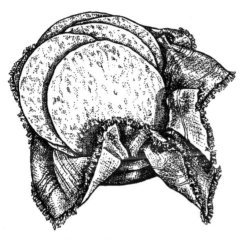

You Will Need:

(8 servings)

4 corn tortillas

4 slices of cheese

To Do:

Wash cooks' hands

Clean a toddler-height table with toddlers' help. Let them watch as the cheese is cut into triangles and placed on the (circle) tortilla. Talk about it. Fold tortilla over the cheese. Heat quesadilla in the microwave or fry pan. Turn in fry pan when cheese begins to melt. You can put a little margarine in the pan if you like.

Resource Book:

Fit-A-Shape, *Shapes*

New Words:

quesadillas (K-sa-deyias)		cheese
tortillas	turn	triangle
circle	fold	hot
melt	bake	

Sample Questions or Comments:

Talk about triangles and circles. Are there other circles or triangles around?

Sometimes it is smart to keep comments other than "yum" or "Doesn't it smell good!" to a minimum. As we have said, toddlers seem to interpret comments as pressure to try something new.

Note:

Try to allow toddlers to watch and smell the quesadillas cook. Talk about the cheese melting. They will be HOT when you take them from the oven. Let them cool—count to some astronomical number while you wait for them to cool. Then cut in 1/2 before serving.

You can add meat or mild salsa with the cheese if you like.

Save for Older Children:

Cutting the cheese and the tortillas, and putting them into the microwave themselves. Four-year-olds love to control the microwave. Talking about Mexico and their great foods.

No Heat Cooking "Space Sticks"

You Need:

1/2 c peanut butter (with no sugar)

1/2 c corn syrup

3/4 c non-fat powdered milk (keep aside a little to coat the space sticks at the end) or use 1/2 c sifted powdered sugar (optional)

Bowl and spoons

Rubber spatula

Shaker with powdered milk or powdered sugar (optional)

To Do:

Let a toddler pour each ingredient into the large bowl after you have ceremoniously measured it, counting each step as you go along. Then stir. Toddlers may not be able to handle the sticky mess until you have stirred it for a while. Then, assuming the "cooks" have washed their hands, they can help knead it. Let a toddler sprinkle the table with a little powdered milk or powdered sugar. Allot some of the dough to each toddler and show her how to knead and roll the "space stick."* After coating each stick in powdered milk, place on a small paper with toddler's name and let dry. Eat at snack time. This dough can also be rolled, cut with cookie cutters & baked at 350°.

New Words:

peanut butter	corn syrup
powdered milk	powdered sugar
sticky/soft	sweet

Sample Questions or Comments:

"Now we will measure the corn syrup. Peter can pour the syrup into the bowl." "How will we get the rest of the syrup from the measuring cup?" "This is a spatula. Jan, will you use the spatula to scrape the rest of the syrup into the bowl?" "Next we will…" "Then we will…"

Save for Older Children:

You can make this dough into balls* and/or roll them in granola, chocolate chips or any other interesting coating you can invent.

*Caution Note:

This dough can be tacky. Sticks are best for toddlers because balls might go into a mouth all-at-one-time and choke toddler.

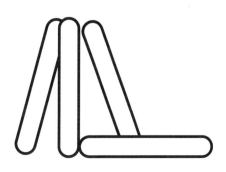

Tasting Sweet & Sour Food

You Will Need:

Lemons (1 lemon yields about 2 to 3 Tbsp of juice)

Sugar to taste (most recipes call for a whopping 3-4 Tbls per lemon!), try for less

Water—2 cup per large lemon (try for more water)

Lemon squeezer, sieve, pitcher, spoon and knife (for adult)

Resource Book:

Kathie Billingslea Smith & Victor Crenson's. *Tasting*

Tip:

The lemonade flavor is enhanced by boiling a small amount of the water with the sugar and then using the syrup to make the drink. This step loses the immediacy of the experience, however.

To Do:

Let toddlers look at and touch the lemons and see the sugar. Have cooks wash hands and the lemons. Cut the lemons and have toddlers taste them, and then taste the sugar. Talk about the difference. Let them squeeze the lemons and then together, make lemonade. Let them watch the sugar disappear. Give them a taste and decide if there is enough sugar. Then taste when it is just right. Refrigerate before snack time if possible. No real need for ice for toddlers.

New Words:

lemon, lemonade   yellow

sugar   sour, sweet, "just right"

Sample Comments or Questions:

"Can you guess what these are?" "Right, they are lemons. Lemons are a pretty yellow—the same color as Jason's shirt." "I wonder how a lemon tastes?" "My, they are sour, aren't they?" "Some people like to eat a lemon like this." "I like a little sugar on my lemon." "How do you suppose the sugar will taste? Let's taste the sugar." "Now, I will pour the sugar into a little water and stir. Oh, where did the sugar go?"

Note:

Toddlers learn most when they can touch, feel and taste something. Have enough thin lemon slices for each toddler to taste one. You can cut more to make the lemonade if necessary.

Save for Older Children:

Talking about lemon as a citrus fruit and that it is related to the orange and grapefruit. Comparing sweet things like sugar, candy, jam, prunes (sweetmeats), cookies with each other.

Appendixes

BIBLIOGRAPHIES

Adult Resources

Early Childhood Books

Cassettes, CDs, Videos and Toys

AGES AND STAGES

HOMESCHOOLER RESOURCES

Adult Resource Books

N=Nutrition

If you are a teacher needing to include science in your daily curriculum, you will want to know about the American Association for the Advancement of Science (AAAS) reference series: *Science Books & Films' Best Books for Children*. It is an annotated bibliography listed by subject with lots of cross references. Science books are no longer boring. The books the AAAS reviews are as varied and diverse as life itself. Children's' books on science will delight and inform young minds. The series is probably available in your library as reference books, or on microfilm.

N   Barrett-Dragan, Patricia and Rosemary Dalton. *The Kid's Cookbook: Yum! I Eat It.* Concord, CA: Nitty Gritty Books, 1982.

Some easy and nutritious recipes. For older kids but full of ideas for easy cooking fun.

Baskwill, J. *Parents and Teachers: Partners in Learnin.,* NY: Scholastic, 1989.

Begley, Sharon. "Your Child's Brain: How Kids Are Wired for Music, Math & Emotions." *Newsweek,* Feb 19, 1996. pp. 55-62.

Fascinating article.

N   Bowes & Church. *Food Values of Portions Commonly Used* by Jean Pennington. NY: Lippincott-Raven, 1993.

Castle, Kathryn. *The Infant & Toddler Handbook.* Atlanta, GA: Humanics, Rev. 1994.

Coldrey, Jennifer. *Discovering Slugs and Snails* (Discovering Nature Book). NY: Bookwright Press, 1987.

Aid in getting a terrarium set up to collect slugs and/or snails.

Cornell, Joseph. *Sharing Nature With Children.* Nevada City, CA: Dawn Publications, c1979.

Sensitive and full of ideas. Lots of activities for older children.

N   Cryer, Debby, Thelma Harms and Adele Ray. *Nutrition Activities for Preschooler.* NY: Addison Wesley, 1996.

Most activities too old for toddlers, but the simple activities are related to art, blocks, science and math as well as pretend play. Recipes and checklists are helpful. Helps implement a nutrition program.

Darling, Dr. David. *Spiderwebs to Skyscrapers: The Science of Structures.* NY: Dillon Press, c1991.

Helpful for the adult who gets questions about how things are built—geodesic domes.

Discovery Toys. Toys and books are rated by cognitive skills, e.g. "Thinking— Cause and Effect, Forming & Understanding Concepts, Problem Solving, Logic, Reasoning and Memory." One example is *When?: a Discovery Toys Question Book.* Ottenheimer Pubs., c1985. Discovery Toys also offer learning software for children age 3 and up, including *Millie's Math House* that is delightful as well as educational. Discovery Toys are available by catalog through telephone white pages.

Dorling Kindersley, Ltd. *Incredible Comparisons.* NY. 1996.

Beautiful graphics compare length, height, width, depth, speed, etc. of natural & man made objects.

My First Look at Nature. NY: Random House, 1991.

Vibrant illustrations divided into seven groups flowers, minibeasts, trees, animals, water life.

N   Drake University Head Start. *We Made it Together.* Des Moines, IA. c1994.

This delightful fund raising cookbook is full of ideas and has simple pictures illustrating each recipe including "Wash hands."

Duff, Eleanor, et al. *Building Successful Parent-Teacher Partnerships*. Atlanta, GA: Humanics.

Ideas and solutions.

Encyclopaedia Britannia.

A most thorough and accurate resource available in book or software forms.

Ehrlich, Robert, *Why Toast Lands Jelly-Side Down*, Princeton University Press, 1997.

Intentionally uses "low tech" and inexpensive materials. Key principles of physics made surprisingly easy.

Eyewitness Books. NY: Knopf.

Elegant hardcover book series of over 40 subjects including: *Mummy, Knights, Plants, Fossils, Shells, Arms and Armor, Bats, Birds of Prey* and *Amphibians*. Intended for older children, but a 5-year-old I know loves the book on sharks. Adults will cherish them as reference books when the questions come. Many beautiful photographs. Perfect gifts. Eyewitness Books, as well as Eyewitness Explorers (for young readers) and Eyewitness Juniors (paperback) are truly international products published under the DK Logo, in London (Dorling Kindersley) and New York (Knopf) and produced in places like Singapore and Verona, Italy.

Fornari, Guiliano. *Inside the Body*. NY: Dorling Kindersley, 1996.

See the body's inner workings by peeling back the "skin."

N *Food, Family & Fun: a Seasonal Guide to Healthy Eating*. Washington, D.C.: US GPO, 1996.

Green, Jen. *Why Throw it Away?: Making Mad Machines*. NY: Gloucester Press, c1992.

Ideas for crafts that work. Great instructions and good photographs.

N Gregory, Valiska. *Riddle Soup*. NY: Macmillan Inc. 1987.

Good reference for ideas on nutrition and how to stimulate children's interest in trying new foods. See also pamphlet Nutritious Books for Young Readers**

Hefner, Christine Roots and Kathryn Roots Lewis. *Literature-based Science: Children's Books and Activities to Enrich the K-5 Curriculum*. Phoenix, AZ: Oryx Press, 1995.

Lots of ideas that could be scaled down for toddlers.

Hirsch, Elisabeth S., ed. *The Block Book*. Washington, D.C.: National Association for the Education of Young Children. 3rd.ed. 1996.

Still a Fascinating text and chart describing the role of block play in learning about science, mathematics, social development, physical development, art, languages arts as well as working with others. Available through their catalog.

Kamii, Constance. *Number in Preschool & Kindergarten: Implications of Piaget's Theory*. NAEYC #103. 1982.

Young Children Reinvent Arithmetic: Implications of Piaget's Theory. NY: Teachers College Press, 1985.

Ms Kamii's research and writings are well respected in the field of Early Childhood Education. Children learn about number by voting, dividing snacks, playing games, even cleanup.

Kohl, MaryAnn and Cindy Gainer. *MathArts: Exploring Math Through Art for 3 to 6 Year Olds*. NY: Gryphon House, 1996.

Most included activities are too old for toddlers, but you can start toddlers on some of the basic concepts, and what better way than through the arts.

Ladybug. Peru, IL: Carus Corp.

Bright prize-winning younger version of their Cricket magazine.

N Lasky, Vicki. *Feed Me I'm Yours*. NY: Meadowbrook Press, c1994. Distributed by Simon & Schuster.

Fun and full of ideas of

foods children will eat.

N *The Taming of the C.A.N.D.Y. Monster.* Wayzata, MN: Meadowbrook Press, c1978.

N Lember, Barbara. *A Book of Fruit.* NY: Ticknor & Fields Books for Young Readers, 1994.

Great photos of a peach, pear, etc. and where the fruit grows, e.g. a cranberry bog.

Oppenheim, Joanne, and Stephanie Oppenheim. *The Best Toys, Books, and Videos for Kids: The 1995 Guide to 1,000+ Kid-Tested, Classic and New Products for ages 0-10.* NY: Harper Perennial, 1994+.

Useful, well organized listings of toys, books and videos that will help adults who work with toddlers find age appropriate resources. It even describes not-so-appropriate ones at each age. Their description ("What to Expect Developmentally") of toddlers is especially good. This book is updated regularly.

N *Nutrition Action Healthletter.* Washington, D.C. Monthly.

Useful regular updates of nutrition/health information. Recommended by people like Jane Brody.

Ossermann, Robert. *Poetry of the Universe.* NY: Anchor Books Doubleday, c1995.

For your own fun. *NY Times* says, "(Osserman)

has drawn a picture of the universe that is at once magnificent and truly charming."

de Paola, Tomie. *The Cloud Book.* NY: Holiday House, c1975.

This delightfully illustrated book for young children, actually, is full of information that adults will find useful. Do you know the difference between cirrus, cumulus, and stratus clouds? This book is still available in your public library.

Parents' Choice. Quarterly Periodical. P.O. Box 185, Waban, MA, 02168.

Treat yourself to this "consumer guide to children's books/toys/video/audio/computer programs/TV/Rock & pop."

N Parents' Guide. *(Cholesterol in Children): Healthy Eating is A Family Affair.* Washington, D.C.: US GPO, 1992.

Readleaf, Rhoda. *Open the Door Let's Explore More!: Field Trips of Discovery for Young Children.* Readleaf Press, 1996.

Filled with new ideas for field trips. Some ideas are too old for toddlers, but could trigger other ideas that are appropriate.

✔Rowen, Thomas and Barbara Bourne. *Thinking like a Mathematician.* Portsmouth, NH: Heinemann, 1994.

Adults and children learn-

ing together. New ways of teaching math. Building foundations.

Schrank, Rita. *Toddlers Learn by Doing.* Atlanta, GA: Humanics, c1984.

Shermer, Michael. *Teach Your Child Science: Making Science Fun for the Both of You.* Chicago: Lowell House, c1989.

Although I like the idea of "helping learn" rather than "teaching" science, this helpful book is full of interesting ideas and good questions. No pictures.

Simon, Seymour. *Science Dictionary.* NY: Harper Collins, c1994

A wonderful gift to yourself. Over 2000 entries. "Age 9 and up."

N *Tufts University Diet & Nutrition Letter.* Boulder, CO. 12 issues per year.

Usborne Complete First Book of Nature. London: Usborne House, c1990.

Delightful book, myriad of illustrations and full of information. Available in speciality stores.

Visual Encyclopedia of Science. NY: Kingfisher, c1994.

Colorful and helpful for older children and adults. Fun to read with a 10-year-old. Sections include "Planet Earth," "Living World," "Stars and Planets," "Science and Technology." Nice photos, drawings and charts.

Young Children. National Association for the Education of Young Children, 1509 16th St.NW, Washington, D.C. Monthly professional journal.

The NAEYC is an early childhood association dedicated to the support and education of teachers, care providers and administrators who serve children from birth through age 8. They produce a number of excellent books, pamphlets and posters as well as the *Early Childhood Research Quarterly,* that publishes significant early childhood research and scholarly information. Sample Article: Wilson, Ruth. "Nature and Young Children: A Natural Connection." *YC,* Sept. 1995. pp. 4-11.

A little scholarly but full of ideas for opportunities to learn outdoors.

Books to "Grow Into"

The following are some of the most intriguing and colorful
early childhood science and math books available in bookstores and libraries.
Many can be used directly with a toddler (sometimes "talking the pictures")
and others can be resources for you for now.
The toddler will grow into them before you know it.

Toddler Books

Aliki. *My Five Senses*. Stage 1, Let's Read and Find out Science series. NY: HarperCollins. c1989.

I'm Growing. Stage 1, Let's Read and Find out Science series. NY: HarperCollins, c1992.

Two of a wonderful series of dozens of inexpensive colorful science books that also includes titles like:*My Feet, My Hands, Fireflies in the Night, Feel the Wind* ✓ and *Digging Up Dinosaurs*.

Baby Animals Series. Hauppange, NY: Barron's. Including Dodds, Dayle Ann. *Wheel Away!* Illustrated by Thacher Hurd. NY: Harper and Row, c1989.
Colorful. Fun with words. Wheels & circles.

N   Chubby English/Spanish Books. *Let's Eat; Vamos a Comer*. NY: Little Simon Merchandise, c1992

Let's Play; Vamos a Jugar. NY: Little Simon Merchandise.

Let's Take a Walk; Vamos a Caminar. NY: Little Simon Merchandise.

What Color? Que Color? NY: Little Simon Merchandise.

Sturdy series of toddler books with large Colorful illustrations.

Brown, Margaret Wise. *Goodnight Moon*. Illus. by Clement Hurd. NY: Harper and Row, 1947.
Deliciously simple. Toddlers love this book.

Carle, Eric. *The Very Hungry Caterpillar*. NY: Putnam Pub. Group, 1981.
A colorful old favorite. Still available in stores and at the library.

The Very Busy Spider. NY: Putnam Pub, c1984.
Now available in a board book. About adding. among other things.

Carter, David. *Feely Bugs: to Touch and Feel*. Santa Monica, CA: Little Simon, c1995.
A "buggy" *Pat the Bunny*.

Feldman, Judy. *Shapes in Nature*. Chicago: Eyewitness Juniors. NY: Knopf, Children's Press, 1991.
A little old for toddler but one of few books that begin to address simple math concepts for young children.

Gome, Taro. *Everyone Poops*. Kane/Miller Book Publishers, c1993.
Useful and fun. Even elephants do it.

Johnson, Angela. *Rain Feet*. NY: Orchard 1994. Board book.

N   Katzen, Mollie and Ann Henderson. *Pretend Soup and Other Real Recipes: A Cookbook for Preschoolers and Up*. Berkeley, CA: Tricycle Press, c1994.
Great fun with many appealing ideas about cooking with children, and watching children cook.

Lorenz Books. *Farm Animals: A Very First Picture Book*. NY: Anness Publishing. c1996.
Excellent photographs. Sturdy book.

McBratney, Sam. *Guess How Much I Love You*. Cambridge, Mass: Candlewick Press, c1995.
Popular book now in board form. Toddler's arm span is a good unit of measure.

McKee, David. *Elmer's Weather*. NY: Morrow, c1994.
Colorful board book. Elmer is an Elephant who encounters rain, fog, snow, lightning and wind.

Oxenbury, Helen. *Clap Hands*. NY: Little Simon.

Parsons, Alexandra. ✓ *Amazing Birds*. NY: Knopf, c1990.
Great photographs, full of information. Books to grow on.

Petty, Kate. *Ducklings*. c1993.
Inexpensive. Great photographs. Watch for *Bears, Chimpanzees, Elephants, Kangaroos, Kittens, Lions,*

Pandas, Ponies and Foals, Puppies, Seals, and *Tigers.* ✓

Rainbow Books. *Opposites.* Lincolnwood, IL: Publications International, c1995.
Colorful, inexpensive board book with lively photographs.

Robinson, Irene and W.W. *At the Seashore.* NY: Macmillan.

Picture Book of Animal Babies. Macmillan.

✓Santoro, Christopher. *Lift a Rock Find a Bug.* A Chunky Flap Book. NY: Random House c1993.
Sturdy, colorful with lots to talk about.

Selsam, Millicent. *Backyard Insects.* NY: Citation Press Book, Scholastic, c1981.
Inexpensive, colorful photographs. The text is too old for toddlers, but the colorful photographs are large enough and appropriate for them.

✓ *Dandelions.*

Snapshot™ Books. *Things That Grow.*
Series of colorful photographic board books with titles like *Shapes Galore* and *Super Senses.*

Tresselt, Alvin. *Follow the Wind.* NY: Lothrop.
Raindrop Splash. NY: Lothrop, c.1946.
Wake Up Farm. NY: Lothrop, c1991.

✓Usborne Books at Home. Board books for babies and toddlers: *Baby Animals; My Body; Who's Making That Noise?* (1995); *Who's Making that Smell?* (1995); *Playdough; World of Wildlife.* NY: Usborne House.
Colorful series. Available through catalog, home shows and a few nature specialty stores. Consultants can be found in the telephone white pages. Corporate offices in Tulsa, Oklahoma.

Waddell, Martin. *Can't You Sleep Little Bear?* London: Discovery Toys (Walker Boheaoks, Ltd.), 1988.

Owl Babies. Cambridge, Mass: Candlewick Press, 1992.
Three baby owls wake to find their mother gone. They are brave and she comes home safely.

Wee Pudgy Board Books. *Wee Wonders of Nature.* NY: Gorso & Dunlap, c1988.
Pretty, little, sturdy and colorful.

Williams, Vera. *More, More, More Said the Baby.* NY: Greenwillow Books, c1990
Now available in paperback. "More" is one of a toddler's first words.

Wood, Audrey. *Quick as a Cricket.* International Child's Play Ltd, 1995.
"Slow as a snail, small as an ant..." A happy book.

Zeifert, Harriet. *.Sam's Boo-Boo* (Toddler Series). NY: Dorling Kindersley Publishing, 1996.
Comes with a "boo-boo bug" icepack.

Books for Early Childhood

Adler, Irving. *Mathematics.* Illustrated by Ron Miller. NY: Doubleday, c1990.

Good introduction to science of math, basic concepts. Math games and activities.

Arnosky, Jim. *Secrets of a Wildlife Watcher: A Beginners Field Guide.* NY: BeechTree Books, c1991.

Useful tips, nice sketches. Information on finding and observing animal tracks from beetles & frogs, to deer.

Atherlay, Sara. *Math in the Bath (and other places too).* NY: Simon & Schuster, c1995.

Fun and colorful.

Barton, Byron. *I Want to Be an Astronaut.* Mexico: Harper Collins, c1988.

Colorful and simple. Counts from 10 to 1.

♪ *Machines at Work: Airplanes, Boats, Trains, Trucks, Airport, Wheels.*

Berger,. Melvin. *Germs Make Me Sick.* NY: HarperCollins, c1995.

Available in a Read-Along book/cassette version.

Bobbie Kalman Books. *Crabapples.* Crabtree Publishers.

Great photos—lovely color.

Butterflies & Moths. 1994.

ν *Frogs & Toads.* 1994.

Bugs & Other Insects. 1994.

Burton, Robert. *Eggs: a Photographic Story of Hatching.* NY: Dorling Kindersley, 1994.

Carlstrom, Nancy White. *Let's Count it Out, Jesse Bear.* 1996

Learning how to add one more to each number. Rhymes are bouncy and fun. Part of the Jesse Bear series.

N Christenberry, Anne & Barbara Stevens. *Can Piaget Cook?* Atlanta, GA: Humanics, 1984.

Children 5-9 enjoy science experiences in food related activities. Some ideas can be used with toddlers.

Clemson, David and Wendy. *Mathematics.* NY: Dorling Kindersley, c1994.

Colorful. Good book to "talk about," maybe even with toddler. Clear good sized pictures of scissors, pen, string, etc.

Cohen, Miriam. *Jim's Dog Muffins.* Greenwillow. 1996.

A little boy's dog dies and his classmates try to cheer him up.

Cole, Joanna. *The Magic School Bus* series. NY: Scholastic.

Episodes from the animated TV series are based on these books. Fun. Busy pages. The images are too small for toddlers, but the ideas are delightful.

Cousins, Lucy. *Katy Cat and Beaky Boo.* Candlewick Press, 1996.

Bright and colorful "concept" book. Counting, patterns, and animal sounds. Flaps are in jeopardy from young toddlers, however.

√ Cornell, Joseph. *Sharing Nature with Children.* Nevada City, CA: Dawn Publications, 1979.

The classic parents' and teachers' nature awareness guidebook. Full of good ideas. Sensitively written. Identifies activities with the child's age (4 and up), and the moods desired (calm, active or energetic).

Dawe, Karen. *The Pond Book.* Workman Publishing.

For 8-year-olds and up about "things that slither and swim."

Dinosaurs. A First Discovery Book. NY: Scholastic Inc. c1991.

Other First Discovery Books include *Bears; Birds; Boats; Cats; Weather; The Tree;* etc.

Sturdy, colorful, and inventive.

Discovery Toys, Pleasant Hill, CA. Question Books answer *Who?, Where?, When?, Why?,* and *How?* Colorful and inexpensive for elementary school children.

Dodds, Dayle Ann. *Wheel Away!* NY: Harper & Row, c1989.

Colorful. Fun with words." Pa-da-rump pa-da-rump pa-da-rump-pump-pump."

N Ehlert, Lois. *Eating the Alphabet; Fruits and Vegetables from A to Z*. NY: Harcourt Brace Jovanovich, 1989.

Bright, clearly labeled with extra information. Reviewers say this book is ideal for teaching the alphabet and categorization concepts. 3-6

Fit-A-Shape. *Shapes; Colors;* and *Opposites*. Quarto Children's Press, Ltd., c1996.

Gibbons, Gail. *Flying*. NY: Holiday House Book., c1986.

Simple, colorful history of flying. "People have always wanted to fly…" Includes balloons, unusual flying machines, Wright Brothers, weather balloons, blimps, hang gliders, pontoons, ski, helicopters. Other Gail Gibbons books: *Tunnel; Boats; Tools; Fire! Fire!; Catch the Wind;* and *All about Kites*.

N Gregory, Valiska. *Riddle Soup*. NY: Macmillan, 1987.

Recommended for age 4-7 years. "Shows parents how guessing games like riddles can reinforce nutrition learning and stimulate children's interest in trying new foods"

N Hayes, Sarah. *Eat up, Gemma*. NY: Lothrop, c1988.

Henwood, Chris. *Earthworms*. NY: Watts, 1988.

Snails and Slugs (Keeping Minibeasts Series). NY: Watts, 1988.

Both books contain many color photographs— includes collecting, maintaining and observing common "minibeasts."

Hines, Anna Grossnickle. *Sky All Around*. NY: Clarion, 1989.

Designed to introduce young readers to the joys of the night sky.

✓ Hoban, Julia. *Amy Loves the Rain*. NY: Harper & Row, 1989.

One of a series of delightful Amy Loves books. She loves the wind and the snow.

N Hoberman, Mary Ann. *Seven Silly Eaters*. NY: Harcourt Brace, 1997.

I can Read About…Spiders. NY: Troll, c1977.

Inexpensive with good drawings and information, but too old for toddlers.

James, Robert. *Nose*. Human Body. Vero Beach, FL: Rourke Press.c1995.

Great photographs. Discusses cartilage, mucus, sense of smell and sense of taste.

N Jeunesse, Gillimard and ✓ Pascale de Bourgoing. *Fruit*. NY: Scholastic First Discovery Books, 1991.

Recommended for age 4-8 years. Colorful to look at , inside pieces, watch seedlings, sprout, etc.

Birds. A First Discovery Book by Scholastic Inc.

This series is special, sturdy and beautiful to see. Over thirty subjects include *Ladybugs, Bears, Boats, Cats, Colors, Dinosaurs* and *The Human Body*. Some books have brightly painted transparent pages.

Kohl, Mary Ann & Jean Potter. *Science Arts.* Fellingham, WA: Bright Ideas for Learning, Bright Ring Pubs, c1993.

Colorful, lots of ideas.

Kohl, Mary Ann and Cindy Gainer. *MathArts.* Beltsville, MD: Gryphon house, c1996.

Kraul, Walter. *Earth Water Fire and Air: Playful Explorations in the Four Elements.* Edinburgh: Florin Books, c1989.

This craft book for six year-olds and up has many projects challenging enough for a skillful twelve-year-old.

Lester, Helen. *Listen, Buddy.* NY: Houghton Mifflin, c1995. Let's Read-and-Find-Out Science Series—Stages I through IV.

A series of dozens of interesting, colorful, inexpensive science books. Pictures are simple enough for toddlers. One author Aliki has written several that are especially appropriate for toddlers. See Picture Books.

Ling, Mary, ed. *In the Sea.* NY: Snapshot Books, c1994.

Great photographs of fish.

Linn, Margot. *A Trip to The Doctor.* NY: Harper & Row, 1988.

A Trip to the Dentist. NY: Harper & Row. 1988.

Martin, Bill Jr. and John Archambault. *Listen to the Rain.* NY: Henry Holt, c1988.

James Endicott's paintings add a delicate touch to this lyrical book.

McBratney, Sam. *Guess How Much I Love You.* Cambridge, MA: Candlewick Press, c1994.

Delightful "measuring" game.

Medearis, Angela. *We Play on a Rainy Day.* NY: Cartwheel Books, c1995.

Fun and colorful. Good ideas of things to do on a rainy day.

Llewellyn, Claire. *Truck & Other Building Machines.* NY: Dorling Kindersley, c1995.

Exceptional photographs. Good Information. Comparisons. See Dorling Kindersley for more fine books.

Martchenko, Michael. *Birdfeeder Banquet.* Toronto: Annick Press, c1990.

Murphy, Stuart. *Ready, Set, Hop.* NY: HarperCollins, c1996.

Colorful and busy but too old for toddlers.

Nayer, Judy. *Sea Creatures at Your Fingertips.* NY: McClanahan, c1992.

Sturdy, colorful board picture books. Good for preschoolers. Series also includes: *Reptiles; Jungle Life;* and *Night Animals.*

N New England Dairy Food Council. *Nutritious Books for Young Readers.* Pamphlet. Boston. c1992.

A useful bibliography of books that are widely available at bookstores. They make learning about nutrition fun and interesting.

Pallotta, Jerry. *The Yucky Reptile Alphabet Book.* Watertown, MA: Charlesbridge, 1989.

Beautifully illustrated.

Question Books. Includes: *When? What? Why? Who? Where?* Discovery Toys. Available by catalog via the telephone white pages.

Questions and Answers About... NY: Kingfisher Books.

Full of interesting information. Great pictures, Confusing, though—too much on page for young toddler:

Forest Animals.

Freshwater Animals.

Ray, R. *Whose Shoes Are These?* NY: Clarion, 1988.

Read and Wonder Series (Age 4 and up). Cambridge, MA: Candlewick Press. Including Judy Hindley's *A Piece of String is a Wonderful Thing,* and Judy Allen's *What is a Wall After All?*

Inexpensive with lots of detail that might confuse toddlers. We loved *All Pigs are Beautiful, My Cat Jack,* and *Think of an Eel.*

Reed, Teresa. *Rain, Rain, Go Away*. NY: Aladdin Paperbacks, Simon & Schuster, c1996.

Rice, Judith. *Those Mean Nasty Dirty Downright Disgusting but Invisible Germs*. St. Paul, MN: Redleaf Press, c1989.

Delightful. Perfect for 3–5 year olds. Toddlers can enjoy the pictures and the idea of mean, dirty germs.

Rockwell, Robert, Robert Williams and Elizabeth Sherwood. *Everybody Has a Body: Science from Head to Toe*. Mt.Rainier, MD: Gryphon House, c1992.

Ross, Catherine Sheldrick. *Circles: Fun Ideas for Getting A-Round in Math*. NY: Addison-Wesley Publ. c1992.

Schwager, Istar. *Active Minds Numbers*. Lincolnwood, Ill. 1992.

Fine colorful photos of fingers, tennis shoes, etc.

N   Seeling, Tina L. *Incredible Edible Science*. NY: Scientific American Books for Young Readers, c1994.

Selsam, Millicent. *Keep Looking!: Hidden Animals*. NY: Macmillan, c1989.

Another sensitive Selsam book with Joyce Hunt's lovely illustrations. The animal index is useful.

N   Sharmat, Mitchell. *Gregory, the Terrible Eater*. NY: Scholastic Inc., c1980.

Sharp, Richard M. and Norma Seymour Simon. *Wet World*. Cambridge, MA: Candlewick Press.

Sloat, Teri. *From One to One Hundred*. NY: A Puffin Unicorn, c1991.

Delightful counting book. Over 2,500 items are "imaginatively arranged in incredible display of color, detail and inventiveness." This is one of the few books with beginning math concepts that are available for young children.

Smith, Kathie Billingslea and Victor Crenson. *Tasting*. Mahawah, NJ: Troll, 1988.

Good basic science. It should be in every children's' library.

Smelling. Mahwah, NJ: Troll, 1988.

Souza, D.M. *What Bit Me?: Creatures Around Us*. Minneapolis: Carolrhoda Books, c1991.

Good pictures and useful information.

Step Ahead™ Products. Golden Books. *Workbooks, Sticker Starters, Mark and See* books. Racine, Wisconsin: Western Publishing, c1993, 1995.

Colorful, inexpensive sequenced series aimed at preschoolers, kindergarten, K-1 through "Grades 3 and up." The preschool "Before I do Math" is especially appropriate.

Swinton, W.E. *Digging for Dinosaurs*. NY: Young Readers Press, c1962

Text is too old, but the drawings are great.

Taylor, Paul D. *Fossil*. NY: Knopf, 1990.

Another beautiful book of eyecatching photographs by Eyewitness Books.

Tester, Sylvia Root. *A Visit to the Zoo*. Chicago: Children's' Press, c1987.

Although the text is for older children, the photos are great for toddlers.

Thaler, Mike. *In the Middle of a Puddle*. NY: Harper & Row, c1988.

Fun with words. Text is too advanced for toddlers.

Toddler Time™; Thinking; Little Explorers; I can Do It!; Numbers; Sorting, etc. NY: McClanahan Book Co., c1994, 1992.

Inexpensive, bright and full of good ideas, but better for older toddlers. Even better for preschoolers. Stickers are hard to tear and sticking is a skill that comes later. The best part is the "Toddler Tip" to help parents integrate the skills taught into their daily routines.

Varley, Susan. *Badger's Parting Gifts*. NY: Lothrop, Lee & Shepard, 1996.

An affirmative look at mourning.

N Watanabe, Shigeo. *What a Good Lunch!* Philomel Books, c1978.

What a messy eater! Fun.

Watts, Barrie. *Caterpillars: It's Easy to Have a Caterpillar*. Morristown, NJ: Silver Burdett, c1980.

Caterpillars. c1989.

A fine classic series written by Watts in the 80's including *Ants; Beetles; Dandelion; Frog; Duck; Ladybugs; Spider's Web*.

Webb, Angela. *Air*. A Talkabout Book. NY: Watts, 1987.

Sand. NY: Watts, 1987.

Leads children to explore activities like blowing bubbles and making mud pies.

Weeks, Sarah. *Follow the Moon*. NY: Harper Collins, c1995.

Nicely illustrated story of a turtle who got confused and the boy who befriended him.

Whayne, Susanne Sautoro. *Butterflies*. A Nutshell Book. Troll Associates. Central Connecticut State University, c1995.

Inexpensive—good pictures for older toddler. Good "talking" book.

Williams, R.A., Rockwell, R.E., and Sherwood.
√ *Mudpies to Magnets: A Preschool Science Curriculum;* and *More Mudpies to Magnets: Science for Young Children*. Mt. Rainier, MD: Gryphon House, c1990.

Full of wonderful ideas. Classics.

Williams, Sue. *I Went Walking*. San Diego: Harcourt Brace. 1990.

Woods, Audrey and Don. *Piggies*. San Diego: Harcourt Brace Jan, c1991.

Piggies get hot, cold, clean, dirty. ALA Award winner book now comes with a musical tape by Carl & Jennifer Shayden.

Wyler, Rose. *Science Fun with Mud and Dirt*. NY: Simon & Schuster, c1986

Cassettes, Videos, CDs & Toys

Baby Einstein. I Think I Can Productions. 28 min. Musical video. Infants+.

Recommended by Parents' Choice.

Big Bag. TV First series for cable from Children's Television Network. Ages 18 mos.–6 years.

Similarity to "Romper Room" because of talking directly to the audience. Parents' Choice winner.

Eyewitness Videos.

Live action photographic 30 minute VHS videos covering over a dozen subjects including insects, amphibians, birds, cats, reptiles, and sharks.

Feierabend, John. *Music for Very Little People,* Book and Cassette.

Nicely illustrated collection of rhymes and folksongs for ages 6 mos–2 years.

Fink, Cathy and Marcy Marxer. *Blanket Full of Dreams.* CD/cassette.

Award winning, soothing word pictures like "Sleep like a little pea/Growing in the garden."

Glazer, Tom. *Music for 1's and 2's.* Cassettes.

More Music for 1's and 2's. Cassette.

Fun. Both cassettes for the very young, include songs like "Where are your Eyes?" and "What does Baby Hear?" .

Hullaballoo Videos. 30 minute VHS.

Covering opposites, shapes, nature, friends, everything grows.

Jump Start Elementary Learning System. *Jump Start for Toddlers, Preschool, Kindergarten, 1st Grade* and *2nd Grade.*

Excellent series on CD-Rom Windows or Mac. Available through department stores and Discovery Toys.

N   Learning Passport Blocks. *Come & Get It!*

Recommended by Parents' Choice to help youngsters learn about food groups and nutrition while stacking blocks. Handsome images of foods.

McGrath, Bob & K. Smithrim. *The Baby Record* and *Songs & Games for Toddlers.* Toronto, Ont.: 1985. Cassettes.

Great Learning Fun.

Marino, Jane and Dorothy F. Houlihan. *Mother Goose Time:.Programs for Babies and Their Caregivers.* NY: H.W. Wilson Co., 1992.

Rediscover the rhythms of language through traditional rhymes, songs and finger plays, including such favorites as "Head Shoulders Knees and Toes," "Clap, Clap, Clap your Hands" and "Five Little Monkeys".

Mattox, C.W. *Shake It to the One You Love the Best.*

Playsongs and lullabies from Black musical traditions. Nashville, TN, 1989.

Moore, Thomas. *Thomas Moore Sings the Family; My Magical World; Songs for the Whole Day;* and *I am Special Just Because I'm Me.* Cassettes. Thomas Moore Records, Suite 1000, 4600 Park Rd., Charlotte, NC 28209.

Familiar, rousing, everyday experiences in rhythmic song.

My First Science Video. Ages 2–8, 6 VHS videos. Available separately with guides.

N   National Association for the Education of Young Children. *Food for Thought: Nutrition & Children.* Video.

Available through their catalog.

National Geographic. *Geokids.* Videos. Ages 2–6.

Live and animated. "Flying, Trying, Honking Around" birds; "Chomping on Bugs, Swimming Sea Slugs and Stuff That Makes Animals Special." Also available for older children—*Really Wild Animals™ Videos.*

Palmer, Hap. *Babysong.* Cassette/guide and compact disc/guide.

More Baby Songs. Cassette/guide and compact disc/guide.

Both Baby Songs are appropriate for ages 1–3. Fun. Adults will find themselves humming the songs like "Come Read a Book," "Rolling, Walking" that are are especially appropriate for our needs.

Raffi. *Raffi in Concert with the Rise and Shine Band.* From a video soundtrack of Raffi in Concert. Lyrics enclosed, c1989. Troubadour Records. Distributed by MCA, Universal City, CA.

Includes such toe-tapping songs as "Everything Grows" and "It's Time to Sing."

The Raffi Singable Songbook. Illustrated by Joyce Yamamoto. NY: Crown Pubs., 1980.

Riding in an Airplane. Hollywood, CA, 1985.

The Second Raffi Songbook. Illustrated by Joyce Yamamoto. NY: Crown Pubs, Inc., 1986.

There's nobody like Raffi!

See How They Grow Video Series. A series of five video tapes—30 minutes each, allowing ages 2–5 to be right there watching young animals as they are born and grow. Good

Sendak, Maurice. *Little Bear.* Nickelodeon TV, (ages 2–5).

Sharon, Lois & Bram. *Horsey, Horsey; Elephant Show.* Toronto, Ont: 1986.

Toddlers at Work. BMG Video. approx.. 35 min. 18 Mos–3 yrs. Recommended by Parents' Choice.

Toddlerific and *Toddlers on Parade.* Cassettes/guides. Fun and movement combined with learning, include songs like "Let's Count the Animals."

Twin Sisters Productions. *Rhythm, Rhyme & Read; Letters and Numbers;* and *Colors and Shapes.* Cassettes and activity Books. Ages 2–6.

Woody's 20 Grow Big Songs. Marjorie Guthrie. Harper Collins. Includes cassette.

Usborne Books at Home Board Books for babies and toddlers. *Children's Songbook (with Music for Piano, Recorder and Guitar).* Available through catalog and some specialty stores.

N Zucchini Brothers. *In Your Garden.* Cassette. Breezy food hints ("Eat all kinds of things"). Parents' Choice Award. Ages 5 and up.

Learning Materials by Catalog and on Television

Although the libraries and book stores are loaded with colorful books about math, science and nutrition for children, some really special books are available only by catalog, in teacher resource centers or educational materials centers. Feel free to shop at these places, or call them for a catalog. They usually welcome the public and their materials are a nice complement to those available elsewhere. Look them up in the yellow pages under School Supplies—Retail.

Instead of fighting television try to get your children to "learn," try taking advantage of the myriad of tv science shows. Watch them together. Even toddlers like animal documentaries, especially if you talk about what is happening. As the children grow older they will discover shows like *Newton's Apple* that is especially good. As with reading, setting the "stage," making everyone comfortable (popping some corn?), and sharing the experience can make any book or any TV show special to a child.

Ages and Stages**

It is important to be realistic about our expectations of toddler's abilities and disposition.

The infant started learning by merely looking, then by reaching and grasping, and then by experimenting with simple cause and effect experiences. The toddler gets up on two feet, and starts exploring places and examining things, and by the age of two toddler is ready to get ready to read.

During those magic 730 days, from the first step of an unsteady non-verbal baby to being an active verbal three year old, dramatic changes take place. A twelve month old is barely on his feet, where a 2 1/2 year old is running circles around us adults.

Here are a few of the characteristics that are typical of each of the major stages during toddlerhood.

We hope this information will help you find age appropriate ways to nurture and strengthen toddler's natural curiosity.

The Twelve Month Old...

can identify body parts,

may understand more than you think,

understands social approval,

enjoys being read to,

will start testing limits,

will explore, bang, or shake objects with hands, as he crawls, stands and starts to walk.

Their growth rate is starting to slow, so interest in food may decrease.

The Thirteen-Fourteen Month Old...

is striving for independence,

is a wanderer,

says "no" often,

lack of vocabulary makes for frustration,

is most interested in large muscle activities,

is less interested in small muscle activities,

likes to imitate adults including imitating simple housework,

recognizes ownership of objects.

The Fifteen Month Old...

loves to throw and dump,

is active, curious and uninhibited,

little interest in food can start a test of wills,

resists confinement and controls,

has a quick temper that does not last long,

is easily diverted,

cannot really understand "being nice" or sharing,

can use spoon, spilling a little,

removes a garment.

The Eighteen Month Old...

needs lots of physical activity, loves to "dance" to music,

may reject help even if he needs it,

his small vocabulary can cause the frustration level to be high,

biting or hitting, and "no" is a favorite word

still does not know what "later" means,

**Each child develops at a different pace, so most descriptions overlap age groups.

has pride in accomplishments—likes to help

can be afraid of the dark.

The Two Year Old...

runs, climbs, balances better,

emotions are less intense,

can follow simple commands,

begins pretend play,

learning language rapidly,

can tolerate some frustration,

starts concern about pleasing others,

If talking well—this can be a period of equilibrium.

The Two and a Half Year Old...

is always on the go, likes to climb,

clings to strict routines,

has strong emotions again—sometimes to extremes,

has difficulty making decisions,

is persistent,

understands cold, tired or hungry.

The Three Year Old...

enjoys a tricycle, stringing simple beads, will take turns,

is more cooperative, usually can give and take,

is more people oriented,

can start to share and make friends,

has an increased interest in language—can be entertained with words,

is still immature and independent and times,

is still "testing" limits,

is especially ready for preschool, play-group or other social experiences.

Homeschooling Resources

Dear Homeschool "Wanta-bees":

If toddler is your first child and you want to get started with homeschooling, here are some ideas to help you.

You already know that parents are important teachers. Homeschooling is not necessarily a program or a curriculum, but an attitude toward the time spent with your child. In fact, before there were preschools, there were parents doing activities that we would call "learning activities." They called it parenting.

Anyone who works with children needs ideas, information and support.

As you face the daily need for ideas to use with your child, we hope the information about science, math and nutrition resources, and our extensive bibliographies will be of use to you.

The first bibliography is for adults to use when the questions come. The second bibliography lists appropriate toddler picture books, and books that toddler soon will "grow-into." Most of the books mentioned are available in at least one of the three libraries near me, so you probably won't need to buy any new books to enhance your science, math or nutrition efforts. Many public libraries have children's' librarians—usually the most delightful of people—who can be of help to you. Libraries no— puters and some even o classes to older children. We were impressed with their toddler packets that include a toy, a cassette tape and the book, so young children can be truly involved in

the stories. You probably have already participated in your local library story hours for little ones.

The children's' library shelves are filled with intriguing magazines like *National Geographic World, Baby Bug, Cricket, Turtle* and fine homeschooling magazines including: *Home Education Magazine* (P.O. Box 1083, Tonasket, WA 98855), *Practical Homeschooling* (P.O. Box 1250, Fenton, MO 63026-1850), *The Teaching Home* (P.O. Box 469069, Escondido, CA 92046-9060), and *Nature Friend Magazine* (P.O. Box 73, Goshen, IN, 46527-0073). A more comprehensive list of resources is in Don Hubbs' *Home Education Resource Guides,* Tempe, AZ, Blue Bird Pub. c1994. Among the many resource people in your area who want to help you are the community churches, your local Adult Education Office, and your County Schools office. Homeschooling organizations all over the nation are helping families network. They are helping members learn about, reach out to, and share resources with other nearby homeschooling families. Most children's' reference librarians ✓ can help you get in touch.

One national group: American Homeschooling Assn., P.O. Box 1125 Republic, WA 99166-1125, phone 509-486-1351, e-mail: AHA online@aol.com

ublication *California Schooler* is another ple of information avail- to homeschoolers. It is blished by the Home nool Assn. of California n Atascadero, CA. You can also look into the white pages of your

telephone book for more information about local homeschooling. A surprising resource— your local telephone book. Take a tour through the book. In addition to the usual white (alphabetical) and yellow (subject) pages, you can find emergency telephone numbers and information about your local community. Some local telephone books include listings of parks and recreation and maps to help find them; zip codes; services for people with disabilities; and schools and government services. The larger the community the more comprehensive the offerings. Two more sources of information are *Parents' Choice,* a quarterly "consumer guide to children's books/toys/video/audio/computer prog/TV/rock & pop" (P.O. Box 185, Waban, MA 02168), and Joanne and Stephanie Oppenheims' *The Best Toys, Books, and Videos for Kids;* an annual Guide to "1,000+ Kid-Tested, Classic and New Products for ages 0-10," NY. Both are regularly published and are a great help to those of us who work with children.

If you need inexpensive educational materials, try sending a self addressed business-size envelope and $2.00 to:

> Freebies for Educators FT
> 1679 Memorial Park Rd, Suite 179,
> Lancaster, SC 29720

Good luck homeschooling your toddlers.

—Rita Schrank

Some Homeschooling Books:

Guterson, David, *Family Matters: Why Homeschooling Makes Sense,* NY, Harcourt Brace, c1992.

Holt, John. *Teach Your Own: A Hopeful Path for Education.*

Lande, Nancy, *Homeschooling: Patchwork of Days.*

Lurie, Jon, *Allison's Story: A Book About Homeschooling,* 1996.

Macaulay, Susan Schaffer, *For the Children's Sake.*

Pride, Mary, *The "New" Book of Home Learning,* Creation House.

Reed, Donn, *The Homeschool Source Book,* 1994.

Sord-White, Shackley, *Survivors Guide to Home Schooling.*

Voetberg, Julie, *I Am a Home Schooler,* Morton Grove, Ill, Whitman & Co. c1995.

Table 1

POTENTIAL CONTRIBUTIONS OF BLOCKS FOR EARLY CHILDHOOD CURRICULUM

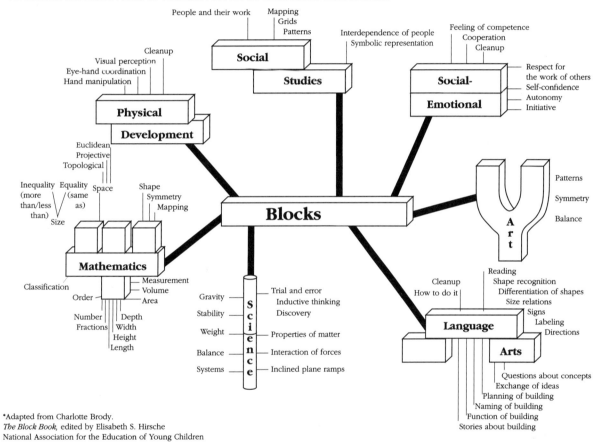

*Adapted from Charlotte Brody.
The Block Book, edited by Elisabeth S. Hirsche
National Association for the Education of Young Children

Index

T

W